EXCHANGE ENTITLEMENT MAPPING

PERSPECTIVES FROM SOCIAL ECONOMICS

Series Editor:
Mark D. White, Professor in the Department of Political Science, Economics, and Philosophy at the College of Staten Island/CUNY

The Perspectives from Social Economics series incorporates an explicit ethical component into contemporary economic discussion of important policy and social issues, drawing on the approaches used by social economists around the world. It also allows social economists to develop their own frameworks and paradigms by exploring the philosophy and methodology of social economics in relation to orthodox and other heterodox approaches to economics. By furthering these goals, this series will expose a wider readership to the scholarship produced by social economists, and thereby promote the more inclusive viewpoints, especially as they concern ethical analyses of economic issues and methods.

Published by Palgrave Macmillan

Accepting the Invisible Hand: Market-Based Approaches to Social-Economic Problems
    Edited by Mark D. White

Consequences of Economic Downturn: Beyond the Usual Economics
    Edited by Martha A. Starr

Alternative Perspectives of a Good Society
    Edited by John Marangos

Exchange Entitlement Mapping: Theory and Evidence
    By Aurélie Charles

Exchange Entitlement Mapping

Theory and Evidence

Aurélie Charles

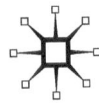

EXCHANGE ENTITLEMENT MAPPING
Copyright © Aurélie Charles, 2012.

All rights reserved.

First published in 2012 by
PALGRAVE MACMILLAN®
in the United States—a division of St. Martin's Press LLC,
175 Fifth Avenue, New York, NY 10010.

Where this book is distributed in the UK, Europe and the rest of the world,
this is by Palgrave Macmillan, a division of Macmillan Publishers Limited,
registered in England, company number 785998, of Houndmills,
Basingstoke, Hampshire RG21 6XS.

Palgrave Macmillan is the global academic imprint of the above companies
and has companies and representatives throughout the world.

Palgrave® and Macmillan® are registered trademarks in the United States,
the United Kingdom, Europe and other countries.

ISBN: 978–0–230–12020–4

Library of Congress Cataloging-in-Publication Data

Charles, Aurélie.
    Exchange entitlement mapping : theory and evidence / Aurélie Charles.
        p. cm.—(Perspectives from social economics)
    Includes index.
    ISBN 978–0–230–12020–4 (hardcover)
    1. Quality of life—Evaluation. 2. Well-being—Evaluation. 3. Distribution
(Probability theory) I. Title.

HN25.C4524 2012
331.01'2—dc23   2011037948

A catalogue record of the book is available from the British Library.

Design by Newgen Imaging Systems (P) Ltd., Chennai, India.

First edition: March 2012

To my family tree.

Contents

List of Figures and Tables   ix

Acknowledgments   xi

Introduction   1

Chapter 1
Subjectivity in Well-Being   9

Chapter 2
Identity, Norms, and Ideals   33

Chapter 3
Exchange Entitlement Mapping   53

Chapter 4
External Shocks on E-Mapping   77

Chapter 5
Social Entitlements   105

Chapter 6
Economic Entitlements   135

Conclusion   157

Notes   161

Bibliography   167

Index   181

Figures and Tables

Figures

4.1   Consumer Price Index (1969–2009)   88

6.1   Employment trend of line workers at the national level
       by gender (1990–2006)   139

6.2   Nominal wages for *maquiladora* line workers by
       gender (1997–2006)   139

6.3   Log of the gender wage gap in the *maquiladora* industry   148

6.4   Log of productivity and real wages over ten years
       (1996–2006)   152

Tables

5.1   Population movements in border states according to
       GDP growth (percentages)   113

5.2   Growth rates in accidental or violent homicides in
       border states (percentages)   115

5.3   Growth rate in different aspects of health (percentages)   118

5.4   Growth rate in different aspects of household structure
       (percentages)   119

5.5   Employment and unemployment rates by gender
       (2000–2007)   120

5.6   Growth rates in labor force composition per industry
       (1998–2003)   121

5.7   Endowment of resources by gender as perceived
      by the gender   128

5.8   Perceived optimality between spouses   130

6.1   Employment proportion and growth of workers
      per output type and gender in the *maquiladora*
      industry (percentages)   145

6.2   Employment proportion and growth of workers
      per output type in the *maquiladora* industry
      (percentages)   146

6.3   Variance decomposition of productivity, real wage
      per male and female worker   153

6.4   Granger causality/Exogeneity Wald tests   154

Acknowledgments

Previous versions of the chapters in this book have benefited from presentations at UNESCO, Paris (HDCA conference), Radboud University Nijmegen (Capability and Happiness Workshop), Concordia University, Montreal (ASE conference), San Francisco (ASSA meetings), New York (EEA conference), and University of the Basque Country, Bilbao (International Conferences on Developments in Economic Theory and Policy). A number of individuals have contributed to the realization of the fieldwork presented in chapter 5, which could not have been completed without the support, kindness, and generosity of Claudia and Carlos, Nelly and Javier, and Ruben, Juan, and Cesar, together with the individuals interviewed and all those who gave their time to complete the questionnaires. I am grateful to the Society for Latin American Studies, University of Leeds, Royal Economic Society, Association for Social Economics, and Eastern Economic Association for financial support at different stages of this research. Finally, I am most grateful to Giuseppe Fontana, my sunshine, and to Malcolm Sawyer, Martha Starr, and John Davis for influencing the writing of this book. I am solely responsible for any errors remaining.

Introduction

Ideals and beliefs depict in people's imagination the perfect human being one wishes to be. People's intents, upon developing capabilities, are guided by ideals, the ideal human being they wish to become, which often conflicts with the reality they are entitled to. The essence of this book is to highlight social norms as the main entitlement failures that prevent one from becoming such an ideal human being, by showing that social norms are the main obstacles to the development of capabilities. To do so, this book develops and implements an innovative tool, exchange-entitlement mapping, or E-mapping (Sen 1981), in an attempt to contribute to the research on human development and the Capability Approach (CA), which considers poverty as the deprivation of human capabilities. In particular, this tool enables us to look at the economic and social opportunities available for a particular group of individuals sharing a common identity to develop human capabilities. While doing so, one research objective here is to identify the channels through which economic events affect individual well-being and collective well-being.

Over the past decades, research on well-being has been mainly addressed through two approaches, the Capability Approach (CA) and the broadly defined Happiness Approach (HA). Both approaches are quite different and are often considered by commentators as competing views on individual well-being. The end of the CA is for human beings to be able to develop capabilities they have reason to value, in Sen's view, or to develop fundamental capabilities, in Nussbaum's view, with happiness being one of the components of well-being understood as human flourishing. On the contrary, the end of the HA is for human beings to achieve a subjective sense of well-being or overall life satisfaction. Despite promoting different goals to well-being, the methodology adopted by both approaches is to start from the point of view of the individual, which allows some synergies between the two approaches. One of the synergies between the two approaches is related to the issue of subjectivity. In the CA, people chose the capabilities they have reason to value, as in Sen's tradition, or capabilities that are universal, as promoted by Nussbaum's list of fundamental entitlements. The personal value attached to capabilities is strongly influenced by identity-related ideals (such as cultural or religious ideals, for example), while the list of fundamental capabilities is one view of what capabilities should ideally be. In the HA, subjective well-being is gauged according to a set point or a reference point toward which people tend to go back over time, or according to an optimality point in the utilitarian tradition. This point is equally strongly influenced by identity-related ideals through peer comparison and habit formation. The problem of norms and related ideals in both approaches is an issue that has already been discussed in their respective literatures. However, the E-mapping framework put forward in this book seeks to capture social norms through the way social and market interactions influence identity well-being and, therefore, economic outcomes. Identity well-being means that an individual in that framework is understood to be a unique combination of multiple identities.

Exchange-entitlement mapping, or E-mapping for short, represents the set of consumption bundles that the individual faces, any of which can be chosen, given his or her endowments (Sen 1981). In its original version, Sen used an entitlement approach in the context of famines in Bangladesh caused by entitlement failures to food supply, rather than by an overall shortage of food supply. In that context, the entitlement approach was understood as the breakdown of food distribution and access to food rather than as an overall shortage of food supply. In the present context, however, an entitlement approach augmented with capabilities is used to explain the channels through which economic events affect individual well-being. In other words, the approach developed here seeks to explain the entitlement failures of the economic and social environment to provide the opportunities to develop capabilities one would ideally choose to develop. A capability represents the ability to achieve different states of being according to the commodities available in an individual's environment, whereas a functioning is an achievement of an individual. Entitlement failures mean that individuals' freedom to choose is undermined by the reduced number of alternatives from which they can choose. Just as individuals have a set of achieved functionings and endowments, they also have access to potential functionings inherent to the economic and social system in which they live. Inequalities find their roots in entitlement failures and are then translated into the impoverishment of capabilities.

By including functionings into a dynamic approach to Sen's E-mapping (1981), part of the argument developed here is that capabilities are socially constructed and developed through various entitlements and entitlement failures—where entitlements are understood in the sense of "access to"—and themselves depend on the interdependence of opportunity sets. Each individual is born within a specific social, historical, and economic context, which shapes his or her innate capabilities, "genetic potential,"[1] or personal conversion factors of commodities into functionings.[2] Access or failures of access to further capabilities are shaped by this environment. An individual born in an environment offering a wide range of opportunities to develop his or her human capabilities does not necessarily realize that the entitlement or access to those opportunities is a prerequisite to his or her human development. The CA represents an immense step forward to assess human development, and identifying the entitlement failures that prevent human capabilities from being developed is one way to proceed in terms of policy application of the CA. By using a dynamic entitlement approach to identity well-being, the E-mapping framework developed here seeks to show that individuals are entitled to capabilities depending on their multiple identities. In effect, E-mapping aims at finding the channels through which economic policies affect individual well-being, which leads to entitlement failures. Finding the entitlement failures of a group of individuals sharing an identity seems to be a much more straightforward task than collecting information concerning the capabilities of each individual. A group of individuals can be defined according to one of their social identities in terms of job, class, caste, gender, age, ethnic background, and so on. In that sense, starting from the identity's point of view and "climbing" up to the economic policy itself brings new insights to the evaluation of inequalities between groups of individuals. Each individual is thus affected by economic changes through his or her multiple identities.

Just as all individuals have a set of personal capabilities at any point in time and possess natural endowments from birth, they also face limitations in access to opportunities, inherent to the economic and social system in which they live, which prevent them from achieving their ideal set of identities. Each individual is a unique combination of identities with a certain hierarchy of related ideals. The converging ideal of all identities is to sustainably function in its living environment. Failures to be able to behave toward that ideal are considered to be failures of entitlements to this optimal functioning. Individual entitlements or, in other words, access to economic and social opportunities of individual development, are constrained by the interdependence between the choice set of an identity and the choice set of another identity. Thus, by allowing the interaction of capabilities and identities, the aim is to understand the failures leading to the limitation on the development of capabilities over time, whether in terms of economic entitlements or social entitlements.[3] Entitlement failures affect human development opportunities and lead to inequality in the choice set of different identities, which then translates into the impoverishment of capabilities. The concern here is not so much about the conflict of entitlements between individuals, or about the conflict of identities for an individual, which will appear ultimately into a hierarchy of ideals, rather, it is about the norms that set the rules for the interdependence of entitlements and therefore determine this hierarchy of ideals at the personal and interpersonal levels.

To assess the E-mapping framework, the final part of the book looks at a specific identity: the identity of a *maquiladora* line worker in Mexico. The *maquiladora* industry can be described as wholly foreign-owned or Mexican-owned subsidiary plants, mainly on the Mexican border, for the assembly, processing, and finishing of duty-free foreign materials and components into products for export, essentially to the United States. The *maquiladora* industry is an Export-Processing Zone (EPZ) type of industry, which helped to boost the economic development of Mexico from the 1960s in an effort to implement import-substitution industrialization along the Mexican border with the United States. Mexico, and the *maquiladora* industry in particular, is a widely studied region and serves almost as a laboratory for the investigation of the economic, cultural, and social frictions between a developing country (in this case Mexico) and a developed country (the United States). As is often the case in a love-hate relationship between neighboring countries, specific features illustrate the Mexico-US relationship that has its roots in historical tensions. The main military conflict between these two countries goes back to the Mexican-American war between 1846 and 1848, which saw the annexation of the states of Texas, Alta California, and New Mexico by the United States. On the Mexican side, this defeat created an incentive for the government to develop and colonize the northern border in order to avoid further annexation.

Since 1994, different macroeconomic events have hit the *maquiladora* industry: the implementation of the North American Free Trade Agreement (NAFTA) in January 1994, the peso devaluation in December 1994, the US recession in 2000–2001, and the increasing competition from Asia in particular. In November 2001, China joined the WTO, which meant access to wider market opportunities for Mexico and increased pressure on its competitors, including those in the EPZ. In the past two decades, the increasing tension between Mexico and the United States due to the limitation of NAFTA as a trade agreement only—excluding the free movement of labor—is mainly reflected throughout this book in economic terms (trade, investment, labor markets, and the exchange rate) and in subsequent social changes (cultural habits, household structure, mortality causes). In 2006, migration flows from Mexico to the United States remained a recurrent problem, which led the Bush administration to decide on the erection of a wall on the border at the critical points of illegal migration. In that year, however, more than 28 million Mexicans were already established in the United States, mainly in the border states.[4] Although Mexico is currently established as a high middle-income country, with a GDP growth and size equivalent to South Korea in the late 2000s,[5] the main incentive for Mexicans to migrate to the United States is related to the increase in income inequalities and criminality in the home country and the prospect of higher purchasing power in the United States.

The evidence presented looks at the way in which economic events—namely, the implementation of NAFTA and the peso devaluation in 1994, as well as the US recession in 2001—affected the entitlements to resources and potential functionings of *maquiladora* workers within their households, and the labor market for *maquiladora* workers. From an anthropological perspective, the final part of the book evaluates *maquiladora* workers' freedoms offered by the Mexican society in terms of the use of commodities and the development of their capability sets. Evidence from Brazil, Chile, and Mexico, for example, has shown that limited access to employment and insufficient labor income are the two major economic entitlements through which development strategy and poverty alleviation policies can act.[6] The identity of line workers in the *maquiladora* industry is investigated in order to highlight the Mexican-American economic and social frictions from the point of view of interdependent social identities, that is, as a male or female *maquiladora* line worker in the Mexican society. In effect, in order to build a thorough understanding of these frictions, the identity of the *maquiladora* worker needs to be related to his or her other main identities: as a gendered individual, as a household member, and as an employee of a specific industry. Investigating the different identities of *maquiladora* workers allows us to identify the channels through which macroeconomic events affect their individual well-being. The identity of a "*maquiladora* worker" embraces different aspects of an individual's life, which can be understood in terms of a set of identities. For example, a *maquiladora* worker can be a male worker, a father and a son, a consumer of imported goods, and so on; or a female worker, a mother and a daughter, a singer at a local church, and so on. Thus, a macroeconomic policy can have a positive or negative influence on the capabilities of several aspects of an individual's life: as worker, consumer, breadwinner, and so on, and whether male or female. For example, *maquiladora* workers used to be mainly young female workers, but as the industry changed its structure from a labor-intensive industry to a capital-intensive one, the typical worker became more mature, and was male or female.

Starting from the identity's point of view, E-mapping shows the effect of economic events on individual well-being. Social norms articulate the impact of such events on the identity of different groups by setting the rules for the interdependence of opportunity sets. The group identity whose capabilities have the lowest socially perceived value in monetary terms is the most negatively affected in terms of living standards and the development of capabilities. The discussion in this book is articulated in the following way. The first two chapters of the book look at how ideals influence individual well-being. In that sense, the first chapter looks at the importance of subjectivity in the CA and the broadly defined HA, while the second chapter shows that ideals are related to identities, whether cultural, religious, geographic, or historical. Chapters 3 and 4 then develop the framework of E-mapping and show the importance of the interdependence between identity E-mappings in determining economic outcomes. Finally, chapters 5 and 6 provide some evidence from the *maquiladora* worker's identity to illustrate how the approach of identity can highlight the main causes of entitlement failures to human and economic development, in terms of social entitlements, of which the freedom of socialization is one example, and economic entitlements, or, in other words, access to income, job, goods, and services.

Chapter 1

Subjectivity in Well-Being

Since Aristotle's writings on *eudaimonia*, prominent thinkers have been concerned with the idea of defining and measuring human well-being. Over the past decades, this concern has been addressed by two approaches: the Capability Approach (CA) and the broadly defined Happiness Approach (HA). Both these approaches are quite different and often opposed to each other as competing views on individual well-being. The end of the CA (Sen 1985; 1999) is for human beings to be able to develop capabilities they have reason to value, with happiness being one of the components of well-being understood as human flourishing (Dasgupta 2001). On the contrary, the end of the HA is for human beings to achieve a subjective sense of well-being or overall life-satisfaction. Despite promoting different goals to well-being, the individualistic methodology adopted by both approaches, to start from the point of view of the individual, however, allows some synergies between the two approaches. The aim of this chapter is to analyze the role of one of the synergies, that of subjectivity, on both the approaches. Comim (2005) in effect argues that one of the synergies between the CA and HA is the subjective evaluation of well-being, which comes as a natural consequence of the psychological work on HA research and as a consequence of the origins of CA in moral philosophy. This chapter exploits this synergy by showing that the subjective evaluation of well-being and capabilities are dependent upon social norms. The problem of social norms in both approaches is an issue that has already been discussed in their respective literatures, showing notably how social norms lead to problems of adaptation (Clark 2009) or interdependent preferences (Gasper and Van Staveren 2003; Iversen 2003; Rader 1980). However, the E-mapping framework put forward in this book requires a thorough understanding of the ways the problem of perception affects individual well-being, either in terms of the CA or in terms of the HA. The chapter first proposes a nonexhaustive discussion of the CA and the problem of subjectivity in its theoretical developments and empirical applications. Then, starting with the British utilitarian tradition, the HA and its recent developments are presented in relation to the issue of personhood in the HA. Finally, the last section discusses the problems of interdependence of individuals and the adaptability to life circumstances in both approaches; the chapter then concludes with remarks on the role of norms and ideals on human development.

1.1. Capability Approach

The CA is a multidimensional approach to well-being and human development, and it also seeks to understand poverty and inequality within its framework. The origins of the CA follow a long line of thinkers that can be traced back to Aristotle's analysis of "human flourishing" (*eudaimonia*), Smith's analysis of "necessities," Marx's concern for human freedom, Rawls's emphasis on access to primary goods, and, more recently, the Basic Needs Approach (Alkire 2002; Clark 2006; Nussbaum 2000; Sen 1982; 1999). The CA addresses the importance of people's own evaluation of their human capabilities in assessing well-being, how far human development can be understood in terms of capabilities, and it understands poverty and inequality in terms of opportunities or freedom to develop capabilities people value. A capability represents the ability to achieve a state of being, while a functioning is defined as an achievement of a person: what he or she can do, or be, according to the commodities available in his or her living environment (Sen 1999). Thus, poverty is considered as the deprivation of capabilities, while human development is considered in terms of the freedom to develop the capabilities people value. To reduce inequalities, the rationale behind the CA framework is that all individuals should have access to the opportunities to develop the capabilities they choose to develop, or, in Sen's words, have reason to value. It is precisely the value individuals attach to their capabilities that is subject to subjectivity and relies heavily on individual beliefs. Cultural and religious contexts "distort" people's perception of well-being (Clark 2009), but they equally help to shape people's values of what capability they deem worth promoting. The aim of this section is to look further into the problem of subjectivity and beliefs in assessing capabilities, both theoretically and empirically.

Capabilities, as defined by Sen (1985), represent the various combinations of functionings ("beings") the individual can achieve, while a functioning is an achievement of a person: what she or he manages to do or be. A common example used to understand the concept of functioning is the functioning of being adequately nourished, given a bundle of commodities (food) and depending on the ability to be nourished, for example, whether one chooses to go on a hunger strike or decides to avoid hunger (Clark 2006). As noted by Robeyns (2005b, 100), Sen's use of the term "capability" was, in his earlier work,

> synonymous with a capability set, which consists of a combination of potential functionings. Functionings could therefore be either potential or achieved...A person's capability is then equivalent of a person's opportunity set. But many other scholars working within the capability paradigm, including Martha Nussbaum, have labelled these potential functionings "capabilities." In that terminology the capability set consists of a number of capabilities. One does not find this usage of capabilities (as being the individual elements of one person's capability set) in Sen's earlier writings, and in his later writings he uses both uses of the word capability interchangeably.

Since the theoretical framework presented in chapters 3 and 4 of this book draws mainly on Sen's early work of E-mapping (Sen 1981), the use of the term "capabilities" here may appear confusing, as it relates to Sen's early works. To avoid potential confusion in understanding, and in line with Robeyns's theoretical survey of the CA (2005b), the terms used hereafter will be "achieved functionings" and "capabilities": the latter referring to a capability set of potential functionings. In effect, just as the "budget set" in the commodity space represents a person's freedom to buy commodity bundles, each individual has a "capability set" in the functioning space, which reflects the person's freedom to choose from possible livings (Sen 1992).

Subjective and Objective Capabilities

The founders of the CA, Amartya Sen[1] and Martha Nussbaum, together with their followers, seem to be divided between those supporting Sen's vision of the CA and those who favor Nussbaum's view. The debate[2] deals essentially with the issue of listing capabilities, and it is divided between setting up an objective and universal list of capabilities (Nussbaum 2000), thereby letting individuals promote capabilities they value and setting a list of capabilities through a democratic process (Sen 1999), and favoring a combination of both objective and subjective types of approach to list and promote capabilities (Burchardt and Vizard 2011).

Sen never wished to set up a definitive list of capabilities. Diverse reasons were put forward by commentators, including his understanding of capabilities as being context specific. In effect, Sen wishes not to bear the responsibility of setting up a list of capabilities, since it could differ extensively from one culture to another, from one stage of economic development to another, and from one individual to another (Sen 1992). Sen was involved in the construction of the Human Development Index (HDI), but he rapidly discovered its limitations as a universal measure of human development. The HDI is used by the United Nations Development Program (UNDP) and was implemented in the early 1990s as a tool to assess well-being. In the CA tradition, the HDI measures the level of education, health, life expectancy, and GDP per capita of a country as a whole, thus concentrating on the evaluation of inequalities between countries rather than on inequalities between individuals within each country. The level of aggregation of the HDI in effect undermines one of the strengths of the CA: to start at the level of the individual in order to reveal the possible source of inequalities between groups of individuals. However, national Human Development Reports seek to respond to the need to understand inequalities at the national level. Sen's willingness to avoid setting up a list of human capabilities may come from his practical experience in contributing to the HDI, but also, and especially, from his lifelong respect for democratic deliberation (Nussbaum 2003). People should be allowed to settle these matters for themselves, and as long as their voices can be heard, it is a highly desirable democratic principle. Thus, setting up a list would be against the original nature of the CA in the sense that human capabilities are defined and evaluated according to a list and not necessarily as valued by individuals. But how do individuals attach more value to some capabilities than others? Valuing capabilities means that the relative weight attached to each capability is context specific and individual specific. Context-specific valuation would imply that the weight attached to personal capabilities varies from one event to another, depending on its environment—as promoted by ecological rationality (Smith 2003), discussed in the next chapter. For example, the ability to cope with a broken leg in the middle of a desert may have a higher value for the victim than the same ability within a hospital. Entitlement or access to health services is therefore a vital factor in valuing this capability, which determines the probability of survival in two different contexts. Valuing capabilities also means that the relative weight attached to personal capabilities is liable <stet> to a subjective assessment by each individual and is therefore specific to each individual. Problems of adaptation and interdependent preferences then influence this subjective assessment, as put forward by the literature on subjective well-being reviewed in the next section.

Nussbaum (2000) listed ten central human capabilities as fundamental human entitlements. Her list of central human capabilities is composed of the following: life; bodily health; bodily integrity; senses, imagination, and thought; emotions; practical reason; affiliation (having the social bases of self-respect and no humiliation); relation to other species; play; and control over one's environment. Nussbaum's vision of the CA is that "each and every person has an indefeasible entitlement to come up above a threshold on certain key goods...focusing on the least well-off, refusing to permit inequalities that do not raise that person's position" (Nussbaum 2006, 342). The main strength of her approach is thus to provide a list of universal entitlements, or, in other words, an ideal toward which humankind *should* tend to. Following Aristotle's tradition of human flourishing, capabilities are understood to be human rights, which makes her list an account of fundamental entitlements, describing the minimum standard of social justice for all individuals (Nussbaum 2003). This ideal is, however, constrained by a reality of how individuals *should* behave according to their own ideals and beliefs. Nussbaum's list of capabilities makes them particularly relevant in a society with a high level of political freedom. For instance, in a context of restrained political freedom, the central human functional capabilities, "Play, Affiliation, and Control over One's Environment," are limited, while other functional capabilities can appear to become central in such a context; the ability of obedience or resistance and the ability to cope or not cope with danger (courage or fear) are examples. Ideals and beliefs, whether they are religious or humanist, may equally shape the political order of countries. A number of countries with one political order co-exist with numerous religions, where the church and the state are separated in principle, while other countries make explicit use of the word "God" in their constitutions. Similarly, one political order may have evolved as a consequence of certain religious values. For instance, the political economist and sociologist Weber (1958) famously argued that democracy and market freedom in the United States are significantly correlated with the Protestant ethic. However, it could also be argued that the freedom of religion guaranteed by the US Constitution adopted in 1787 is also correlated with democracy and market freedom, given the diversity of ethnic and religious backgrounds that shaped US society. Here, the intention is not to enter the complex and sensitive debate of democracy versus religion, but rather to give a few insights showing that some freedoms may be *perceived* to be more crucial than others according to the beliefs and ideals of a dominant group of individuals in a country-specific context. In the context of religious states, for example, one can argue that religious freedom may be a prerequisite in achieving political freedom: in such a context, religious freedom can be denied, which could eventually lead to the denial of political freedom. In Nussbaum's framework, not experiencing the ability to worship freely leads to constraints on political freedom, and the resulting lack of capabilities is associated with a lack of potential for human development. Constraints on political freedom, or on any freedom, represent entitlement failures for individuals to achieve their full potential of human development. In Sen's framework, these entitlement failures prevent groups of individuals who do not share the dominant ideal from developing the capabilities valued according to their own ideals.

The problem of subjectivity in assessing human capabilities led some commentators to suggest a method of selecting capabilities through a process of participation by the people with whom the study is concerned (Alkire 2005; Robeyns 2005a). Robeyns argues that in small-scale projects, capabilities should be selected using participatory methods. Also, work on larger-scale projects should start "with some brainstorming," and it is important that it "engages with all relevant (academic and nonacademic) literatures, engages with other relevant lists, and opens the draft list up for discussion" (Robeyns 2005a, 210). This type of approach seems to be universally applicable and more in line with Sen's own point of view. Sen (1982; 1999) often highlights the value of a participatory process between all parts of a community for policymaking and as the basis for a democratic and efficient outcome in social welfare. However, such a participatory process remains quite demanding in terms of the amount of information required to select the capabilities. This represents an important limitation of the CA, since such a procedure requires a vast set of information to be collected (Clark 2006). It also raises doubts about the possibility of cross-country comparison if the lists are different from one country to another, but it may be promising to compare groups of individuals sharing a common identity. Finally, in an attempt to tackle the problems of objective versus subjective capabilities, Burchardt and Vizard (2011) propose a capability-based measurement framework for twenty-first-century Britain combining an objective list of core capability from the international human rights institutions and a subjective list based on deliberative consultation. However, it seems that the debate on objectivity versus subjectivity in capability measurement is hiding a deeper divergence between the two approaches, which revolve around the problem of personhood in the CA (van Staveren 2008).

How does the CA define an individual? Nussbaum's approach to personhood is that the individual is considered as an atomistic entity to which a universal list of objective capabilities can be assigned as a set of functionings that this atomistic entity can choose from. In a sense, this reflects her own critique against Utilitarianism, whether from Bentham's view of hedonism or from Singer's view of preference-satisfaction, which "notoriously refuses such insistence on the separateness and inviolability of persons" (Nussbaum 2006, 342). However, individual capabilities can be "collective" capabilities, resulting from a social environment, and related to the interdependence of opportunity sets (Basu 1987). The interdependence of capabilities is more visible in Sen's approach to personhood through his unwillingness to set up a definitive list of capabilities. The value people attach to their capabilities means that an individual values a certain hierarchy of capabilities to which the individual identifies. An individual can therefore be seen as a collection of capabilities (Davis 2011). The interdependence of human capabilities across individuals, coming from their interdependent opportunity sets, has also been identified by several authors as interdependent preferences (Gasper and Van Staveren 2003; Iversen 2003; Rader 1980), as shall be discussed later in this chapter.

Capabilities are to a large extent socially constructed (Comim 2008). In effect, since the CA is concerned with the value individuals attach to their capabilities, if capabilities are valued collectively, individual capabilities can also be collective capabilities. The concept of individual freedom to value one's own capabilities seems to disappear when the individual attaches more value to collective capabilities than to individual capabilities. In such a context, the distinction between individual well-being and collective well-being is not necessarily clear. For instance, Uyan-Semerci (2007) led a survey among Muslim women in poor areas of Istanbul to investigate the self-evaluation of their capabilities. One problem Uyan-Semerci faced was that the women investigated could not easily consider themselves as individual entities, but instead saw themselves rather as part of a cultural group linked to their religious values. A relational framework where people consider themselves as being part of a collective entity rather than an individual entity is related to the issue of "identity," which will be explored in the next chapter. In the meantime, a collective entity is especially relevant in the context of collective beliefs. As rightly pointed out by Nussbaum, social, cultural, and religious conditioning means that people "simply adapt to the low level of living one has come to expect" through adaptive preferences (Nussbaum 2006, 341). However, a low level of living may also be worth promoting if individual beliefs put greater emphasis on other functionings that would be precluded by a higher level of living, spiritual enlightenment in the Buddhist tradition being one such example. The debate around personhood in the CA is therefore tied to the fact that there is often no clear distinction between individual and collective capabilities.

Capabilities in Context: Some Evidence

The evidence on capability measurement helps to provide some answers to the personhood debate in the CA by trying to find a clear-cut distinction between individual and collective capabilities. The best example of measurement of collective capabilities is certainly the Human Development Index (HDI). Collecting national data on education, income, and life expectancy gives a broad picture of a country's ability to foster the human development of its population. Implemented in 1990, the shortcomings of the HDI discussed above led the UNDP to build a new HDI "2.0," which was first published in October 2010. One critique of the old HDI is that it imposes a single view on human development and places value on dimensions of life that do not reflect every culture or even every person's set of values. The new HDI 2.0 has effectively addressed this shortcoming by relaxing its past assumption of perfect substitutability between its scaled indices, that is, education, life expectancy, and income, but the weight attached to the core components of its indices is still unclear (Ravallion 2010). In effect, the weight attached to each core component is imposed on the general index, but it actually reflects the value HDI officials choose to attach to the three dimensions. For example, the HDI looks at life expectancy as a proxy for a "normal length" life, assuming that the longer one's life, the more one values and desires one's life. The higher bound of life expectancy in the HDI has changed from 85 years, in its 1990 version, to the Japanese life expectancy of 83.2 years, in its 2010 version (Ravallion 2010). The Japanese "ideal" of life expectancy is therefore transposed to an ideal for all countries regardless of country-specific contexts—and, given the notoriously high suicide rate in Japan, regardless of the Japanese's own valuation on the core components of well-being. Similarly, let us assume, for example, that Inuit tribes and other Arctic populations have lower life expectancy than other populations due in part to their natural environment, and that, as a consequence, reaching 83.2 years is the exception rather than the norm: does their resulting lower HDI mean a less desirable life? Life expectancy remains a relevant indicator, but it should be used to complement other biological indicators such as morbidity and caloric requirements according to individual needs, with all the difficulties of data collection these indicators represent (Steckel 2008). However, more importantly, each country should be allowed to set the weight relevant to reflect the value they attach to each core dimension.

As one of the most objective attempts to measure well-being, the HDI starts from an arbitrary weight of its core dimensions. Subjectivity is therefore a key element to well-being measurement, which led some scholars to argue for a participatory process in identifying capabilities. Robeyns (2005a) and Clark (2006) in effect advocate the use of "empirical philosophy" to assess individual capabilities. Clark's empirical philosophy is an attempt to "confront abstract concepts of human well-being and development with the values and experiences of the poor" (Clark 2003; 2006, 8). In other words, he is arguing for a subjective evaluation of capabilities by the individuals concerned to inform the theoretical basis of the CA. For example, in his empirical work based in South Africa, Clark investigated the perceptions of well-being among the urban and the rural poor (see Clark 2006; 2003). He found that people mostly valued the following items, without a predefined order, when defining a "good life": jobs, housing, education, income, family and friends, religion, health, food, good clothes, recreation and relaxation, safety, and economic security. Klasen (1997) also used a household survey to assess poverty in South Africa with a comparative study between an income-based measure of poverty and a more complete measure that included health, education, employment, access to services, and perceptions of satisfaction. Clark and Qizilbash (2005) then revisited Klasen's work, lowering the cutoff between the poor and nonpoor using measures of income, accommodation, and consumption. Their point was to assess the strengths and weaknesses of "fuzzy poverty measures." They highlighted the usefulness of fuzzy[3] poverty measures for the advancement of research on poverty indicators, notably when measuring relative poverty rather than absolute poverty. An important issue here is to understand where the threshold stands between poverty and nonpoverty, and between basic capabilities and nonbasic capabilities. Following this set of empirical studies, it appears that the population surveyed needs to be at the center of the well-being measurement process. The complementarities of self-evaluation and objective evaluation, in the sense of absolute and relative data such as income or education, are essential when measuring living standards. While standard evaluation places a person's living conditions in a general ranking in terms of some social standards, self-evaluation shows how the person would judge his standard of living vis-à-vis others' position (Sen 1987, 30). A report from UNICEF (2007) suggests that the UK and the United States, two of the most advanced economies in the world, come at the bottom of a league table for child well-being among 21 industrialized countries. One out of six categories is concerned with children's own sense of well-being. This category was at its lowest value for both countries. Measuring subjective well-being across countries is critical, given the cultural differences in reporting well-being. However, this report suggests that, despite cultural differences and problems of interpersonal comparability, inclusion of the individual self-evaluation of well-being is now an accepted subjective determinant in measuring welfare and poverty.

Zimbabwe finishes at the bottom of the new HDI 2010 table essentially due to its lowest income index (scoring 0.01) compared with all other countries, but looking solely at its education index component of 0.52, 56 countries have a lower schooling index than Zimbabwe's (Ravallion 2010). So how far does income contribute to capabilities? A number of studies concerning the CA have focused on the relationship between income or expenditure and various capabilities (Altman and Lamontagne 2004; Sen 1999; Burchardt 2005). As suggested by Clark (2006), the results of this research show that income and capabilities do not necessarily have a positive relationship. One reason is certainly the wide and eclectic range of human capabilities such as, just as in Nussbaum's list (2000), the ability to enjoy a healthy life, the ability of expressing free speech, freedom to exercise artistic expression, or the ability to have emotions. Indeed, not all human capabilities have a monetary cost to be supported, and the unobserved sign of the relationship between income and capabilities supports this view. Evidence from early twentieth-century New York City underlines the importance of the household economy as a source of socioeconomic well-being and indicates that cultural norms are a determinant of choice over functionings (Altman and Lamontagne 2004). From a historical perspective, they found that immigrants who were of Russian-Polish Jewish origins, despite having significantly lower purchasing power and more crowded households, were significantly better-off, in terms of mortality rates, than other immigrant communities, and notably better-off than southern Italian immigrants. Following the CA, well-being was defined as a measure of functionings achieved through morbidity and mortality rates. As shown by their results, the reasons are related to the Jewish norms of hygiene and food/nutrition, which differed from the Italian norms of the time. By being better nourished (more meat and fish and less alcohol) despite having a lower income than others, and by being in better health due to better hygiene, the Jews gained in productivity in the household-based factory. Given the multidimensional nature of capabilities, the statement "the greater the income, the more capabilities an individual is able to develop" is not necessarily appropriate. Contextual factors have to be taken into account, such as the provision of social services, the degree of political freedom, antidiscriminatory laws, and cultural norms.

As shown by the literature on the CA described here, subjectivity is an essential component of capability measurement from which a major measurement problem arises: the interdependence of capability sets linked to interdependent and adaptive preferences. The literature on subjective well-being now describes their origin and consequences on human behavior.

1.2. Happiness Approach

Utilitarian Tradition

The relationship between the utilitarian tradition in economics and the current HA finds its rationale in J. S. Mill's point about the desirability of happiness: "Mill is careful to state, more clearly than his predecessors, that utilitarianism is a system of ethical hedonism, i.e. that the criterion applied to individual moral action is general happiness not individual interest" (Welch 1987, 772). Nussbaum (2005) argues that J. S. Mill's view of utilitarianism stands between Aristotle's *eudaimonia* of a flourishing human living and Bentham's simpler view of pleasure-happiness as the end of social planning. Mill's view can be roughly summarized as follows: "A life full of ethical and intellectual excellences and activity according to those excellences does not suffice for happiness, if pleasure (however we think about pleasure) is insufficiently present and if too much pain is present" (Nussbaum [2005] on Mill, 180). Mill's personal development, or individual entitlements from his childhood, generated a personal ability to transform two different intellectual traditions into a unique view of utilitarianism. His unique perspective allowed for two traditional views of utilitarianism to function together. In effect, Nussbaum (2005) also enlightens us on the background to such a convergence of theoretical thinking, which lies in Mill's life experience. Mill's education was built around a father's search for excellence and ambitious achievements—reflected notably in his son's tedious work "A System of Logic" (Mill 1956)—and a mother's lack of emotional warmth, as perceived by Mill, and which led him to a period of deep depression in his early twenties. Therefore, his personal utilitarian perspective on human development is linked to his personal development, which allowed him to create his unique perspective. A search for excellence is a quest for a collective ideal on human development, which may well become universal, but pain and pleasure are traditionally perceived to be a personal experience, which creates a problem of interpersonal comparability.

One legacy of utilitarianism is to seek a universal cardinal measure of well-being, which also creates a problem of interpersonal comparability. Another legacy of utilitarianism is to have contributed to our current knowledge of ethics. Modern philosophical utilitarianism identifies two types of utilitarianism: rule-utilitarianism and act-utilitarianism (Hare 1976; Welch 1987). Act-utilitarianism is related to consequentialism, whereby the rightness of an action is judged according to its consequences: whether or not the consequence is judged to be better or worse than an alternative one. Rule-utilitarianism, however, is about assessing "the rightness of an action by asking whether it would have good consequences if it became part of general practice" (Welch 1987, 774). Harsanyi (1977) argues in favor of rule-utilitarianism, which would make the society better-off as long as individual behaviors are constrained by moral rights and obligations that should not be violated. However, a prerequisite to rule-utilitarianism is to be able to agree on a universal system of moral rights. International organizations such as the United Nations, Greenpeace, or WWF have their own view on what ethics should be about, but context-specific norms of behavior often render their agendas not easily enforceable. As argued by Singer (2002), well-being in the utilitarian hedonistic tradition is a matter of preferences over a set of opportunities, which could help the CA find a definition of the "good," that is, what capabilities are deemed worth valuing and developing.

These preferences over a set of opportunities are, however, constrained by norms or appropriate rules of behavior, which themselves rely on ideals and beliefs shared by individuals.

Harsanyi's preference for rule-utilitarianism is also shown by his critical view of hedonistic utilitarianism, a closer version of utilitarianism to the HA: "It is by no means obvious that all we do we do only in order to attain pleasure and to avoid pain" (1977, 54). However, most psychologists now agree on the fact that "we are attracted to the favorable elements and seek to have them or to prolong them; and we are repelled by the unfavorable elements and seek to avoid them or try to bring them to an end. Psychologists call this 'approach and avoidance'" (Layard 2003, 11). Layard gives no further precision to the meaning of favorable elements. An interpretation of favorable element could imply an element to which one is attracted, which then becomes a tautology. Given Layard's definition of happiness, favorable elements must be the elements that represent a source of happiness, unlike unfavorable elements, which bring unhappy feelings. The psychology literature should bring more light upon our understanding of happiness. The contribution of psychology to the measurement of subjective well-being is notably reflected by the lifetime work of Daniel Kahneman.[4] Most works in psychology strike "the psychological assumptions that utilitarianism must make [which] are so narrow and implausible as to render the theory either inadequate, or positively pernicious" (Welch 1987, 775). The contribution of psychology to well-being assessment is precisely the area in which the debate about the HA is based.

Individual Happiness

Several studies have pointed out that an individual facing a choice will seek the happiest outcome, and that happiness should therefore be the only measure of well-being. Kahneman (2000) and Layard (2003) highlight the fact that human beings seek the personally happiest outcome in any event, that is, the individual estimation of the happiness that will result from a decision prior to taking that decision. In other words, when facing a choice between two alternative outcomes, individuals seek the most immediate source of happiness or the least miserable outcome, regardless of whether it involves acting altruistically or individually. According to Rayo and Becker's model of happiness (2005), the happiness function depends extensively on whether current "success" exceeds a given reference point or threshold. Assuming that happiness is an evolutionary process, their view is that only surprises, that is, unexpected changes, affect happiness by moving away from this threshold. The impact of surprise decreases marginally as a result of adaptation to life circumstances and peer comparisons. However, considering happiness as an evolutionary process, rather than a state of being at a point in time, provides it with a deterministic feature, which would mean that the threshold never evolves through the lifetime. Additionally, it would also mean that happiness can be considered as a single variable, where "negative feelings are simply the negative end of positive feelings," as supported by Layard (Layard 2003, 11). Considering happiness as a single variable, as with Bentham's vision of utility, can be too simplistic, and it can neglect the fact that, as argued by Keynes and Nussbaum, happiness can coexist with pain (Carabelli and Cedrini 2009). Some people may be happy in some domains of life, such as work, and unhappy in others, such as in their marriage/partnership, or vice versa.

One debate within the HA could be described as a fight between aggregated figures of happiness related to life circumstances, coming from the background in economics of most scholars supporting this view—Richard Easterlin being an example—and subjective figures of happiness related to a mental state, relying mainly on Kahneman's lifetime contribution and his background in psychology. Easterlin's research (2005) started from the fact that subjective well-being in Western societies has not, in general, changed much in the past 50 years, while technological and material progress has been tremendous over that same period. Then, using self-reported figures for subjective well-being in different domains of life, Easterlin collected findings relating to income versus happiness, happiness vis-à-vis life stages, country-specific reports, cross-country or cross-community analyses, and so on. As a result, he suggested a theory of happiness based on both psychology and economics where people value different domains of their lives—such as work, family, health, spirituality, and so on—differently. Life circumstances such as marriage, serious injuries, death of relatives, or job loss "deflect a person above or below this set point,[5] but in time hedonic adaptation will return an individual to the initial position" (Easterlin 2005, 29). However, drawing on both psychological and economic empirical studies, Easterlin demonstrated that well-being depends on a variety of pecuniary and nonpecuniary conditions or domains. For instance, the death of a close relative can affect an individual's happiness without necessarily returning to this "set point" influenced by genetic heritage and education. Easterlin (2005) further shows that domains such as family and health have a potential for increasing happiness, which has lost the attention of modern societies. As mentioned above, peer comparison and adaptation to life circumstances are two major factors in determining happiness if happiness is considered as an evolutionary process (Rayo and Becker 2005). Easterlin (2005) suggests that the potential to raise life satisfaction lies in areas in which public scrutiny and, therefore, peer comparison behaviors are minimal. Peer comparison is minimal in the domains of life away from market competition, such as family and health, unlike domains such as jobs or commodities, which provide a social status. For this reason, Easterlin (2005) argues that family circumstances and health improvements have a potential to increase people's happiness that is undermined by market economies, whether it is in terms of jobs through the labor market or consumption of commodities through the market for goods and services. While it can be intuitive that a family focus—without domestic violence—can bring greater happiness, the issue of health as a source of happiness is less straightforward. The inclusion of health as an important domain comes from the empirical evidence as argued by Easterlin: poor health brings low levels of happiness, while improving people's state of health improves their levels of happiness. The effect seems to be more significant if people have experienced poorer states of health in the past. In effect, an individual's happiness tends to be definitely affected when serious health problems or injuries occur. People tend not to return to the set point mentioned above when they have previously enjoyed the highest levels of health. His conclusion is that domains with less public scrutiny and social comparison, such as family and health, leave room for increasing happiness, as opposed to the material and publicly exposed domains, such as goods or work. From Easterlin's perspective, attainment in each domain is not the key to happiness, but rather the relative value attached to all domains. He further argues that the constraint placed on this relative value depends on two types of behavior based on adaptive and interdependent preferences discussed further below.

However, empirical findings in psychology suggest that life circumstances make only a small contribution to the variance of happiness between individuals, far smaller than the contribution of inherited temperament or personality (Kahneman 2000). A person's subjective assessment of well-being is to a large extent a personality trait (Kahneman 2000; Kahneman and Krueger 2006). People's subjective evaluation is crucial in evaluating personal well-being, which has led to a revolution in economics in the past two decades (Frey 2008). Most studies in psychology relate to happiness and well-being essentially as a measure of subjective well-being from the point of view of the individual (Diener and Suh 2000; Kahneman et al. 1999). In all cases, subjective well-being refers to the self-evaluation of people's lives, whether it is in positive or negative terms. Subjective well-being has been a lifelong research interest of Kahneman. Some of his contributions concern the distinction between experienced utility, related to Bentham's concept of utility, or the affect experienced by an individual at time t, and remembered utility at time $t + 1$, which is susceptible to systematic bias in comparison with experienced utility. In turn, individual choice is affected by remembered utility and not by experienced utility or actual experience of utility. Based on laboratory experiments, the evidence provided by Kahneman and Krueger (2006), however, shows that life circumstances have a greater impact on net affect (or experienced utility at time t) than on life satisfaction (or remembered utility at time $t + 1$), which is subject to bias of judgment and perception. The biases are due to shifts in the standards of life satisfaction brought about by peer comparisons. The bottom line of Kahneman's argument is that social welfare should be concerned with the minimization of misery, identified via subjective well-being, rather than with maximizing a vague concept of happiness. In other words, a practical approach to social welfare would be to identify the entitlement failures leading to a physical and mental state of misery. If perception is biased by some prior belief of how things should ideally be, then this perception may lead to a failure of entitlement to capability sets different from what they should ideally be.

Research on well-being and happiness has become rich in the past decades owing to the improvement of data collected through experiments and with the help of neuroscience. Research on subjective well-being has found that there is a high correlation between subjects' self-reports of happiness and the readings of their brains with an electroencephalograph. Evidence suggests that self-reporting and brain readings provide identical results in the measurement of subjective well-being. For example, when shown pictures of a deformed baby in pain and a healthy baby smiling, the self-reported measure of respondents' positive feelings was identical to stimuli in the left part of the brain, while negative feelings were consistently related to stimuli in the right part (Layard 2003). Through MRI scans for brain readings, a neuroscientist at Wisconsin found that positive feelings appeared in the left part of the front cortex of the brain, while negative feelings appeared on the right (Davidson 2000). The self-reported experienced utility of the interviewees was also perfectly matched with the readings of electrical signals in the brain. Based on these results, it is inferred that the feelings of "approach and pleasure" are related to the left part of the cortex while "avoidance and aversive" stimuli are related to the right part (Kahneman and Krueger 2006; Urry et al. 2004). However, other evidence suggests that remembered utility is correlated with brain readings of only 0.30 when using survey reports of life satisfaction (Urry et al. 2004). This means that reporting at time $t+1$ a level of utility experienced at time t

gives inaccurate or biased figures. Such findings support the view that interviews investigating levels of self-reported satisfaction in different domains of life could accurately measure remembered utility biased by a perception of how things should be, and not experienced utility on how things actually are.

1.3. Interdependence and Adaptation

Two features of human behavior have been identified to be highly influential in determining perception of well-being in both psychology and economics, namely, adaptation to life circumstances and the interdependence of individuals. First, a feature of human behavior highlighted by the literature relates to the concept of interdependent preferences in economics, or social/peer comparison in psychology. Interdependent preferences mean that an individual's utility, coming from the ownership of a certain amount of goods, depends partly on the amount of goods owned and enjoyed by others. In other words, the utility function of one individual is related to the utility functions of others. Peer comparison means that individuals compare themselves with other individuals with whom they share a sense of belonging to a common identity. In other words, a group of individuals sharing a common identity will not only help individual members to define themselves as individuals but, adopting the norms of behavior of the group, will also sustain their common identity. Nussbaum's own insistence on the separateness of individuals overlooks the importance of interdependence between individuals in shaping capabilities and functionings. Some functionings would not exist without their interaction with others' functionings. Examples include functionings related to commitment, sympathy, trust, and even sexual intercourse. The individual's entitlement to a network of social values and norms influences the personal development of capabilities by restricting the choice set over potential functionings. In turn, people tend to value their achieved functionings according to the social norms and ideals by which they live.

A second feature of human behavior influencing the perception of well-being is human adaptability to life circumstances. From the happiness perspective, the "endowment effect" (Huck et al. 2005) or "treadmill effect" (Kahneman 2000) represents the notion that people have a tendency to get used to their endowments, in the sense that they adapt their perceptions and judgments to the items they acquire and have access to. Given that individuals seek the happiest events possible, this leads individuals to have hedonic adaptation toward life circumstances (Easterlin 2005). Hedonic adaptation, habit formation, or adaptive aspirations are related to risk aversion, whereby the dislike of loss is greater than the utility from gain. For example, Burchardt's research (2005) suggests that adaptive preferences influence satisfaction levels in the context of constant or increasing incomes but not in the context of decreasing incomes. Rabin (2001), however, argues that risk-aversion theories are intrinsically related to utility derived from wealth, which is a limited proxy for measuring subjective well-being. March (1988) in effect shows that risk preferences depend on the values of possible wealth outcomes in decision-making relative to aspiration levels. As a survival strategy, risk-averse behavior results from adaptation focused on specific targets, which also increases the long-run likelihood of survival. Although this analysis is limited to wealth outcomes, the results show the importance of targets or aspiration levels to be part of a survival strategy.

From the capability perspective, thinking in terms of utility or wealth undermines the multidimensional space and interpersonal comparability brought by the CA. Nussbaum herself mentions the "dangerous" role of adaptive preferences in assessing well-being and in promoting human development, in the sense that individuals can also get used to poor states of living (Nussbaum 2000). Clark (2009), however, is dubious on the role played by the problem of adaptation on capabilities, at least as presented in the current empirical literature. Adaptation or social conditioning of cultural and religious context distorts perception, leading to disparities in the quality of life, even within the household (Clark 2009). Adaptation is, however, a part of human ability to cope with the surrounding environment, as put forward by ecological rationality (Smith 2003). Cultural and biological adaptation to an environment is in itself a human functioning and cannot be ignored by an approach claiming to assess well-being. The problem of adaptation to life circumstances from both the happiness and capabilities perspective is essentially with regard to the reference point to which individuals wish to adapt. In terms of the happiness perspective, the set point to which happiness goes back evolves over time. The literature on identity explored in the next chapter will show how this set point/target/reference point/ideal can be affected by the dynamics of identity. In terms of the capability perspective, choosing from a capability set, a human development path tends to be toward an idealistic view of what a human being should be and thus represent capabilities a person has reason to value. Cultural and religious beliefs go beyond cultural and religious norms. A universal list of human capabilities in itself sets up a belief of an "ideal" state of human being putting forward the freedom of choice.

Norms of behavior influence individual choice over the best immediate outcome in the course of an action. By choosing the least bad outcome of any event according to social norms, the individual can end up in a situation that cannot be fully assessed in positive terms of utility, desire-fulfillment, or satisfaction only, or in terms of capabilities only. To illustrate, let us assume that in the event of being attacked for money, a set of choices faced by the victim includes running, giving up the wallet, or being killed. If the victim is an 80-year-old woman, then her functionings are limited, and the "running" option is ruled out. The utility approach would suggest that, provided she is not killed, the temporary income effect means that she can achieve a lower indifference curve due to the bundle of goods she is no longer able to consume. The CA would suggest that several capabilities have been affected that have lowered her well-being, depending on the value she attaches to her capabilities, and that her capability set has been negatively affected. The affected capabilities range from the ability to buy food, to the ability to live in a secure environment, or the damaged capability resulting from the moral prejudice she has experienced. An approach through entitlements would suggest that, given her limited capabilities and depending on the dominant social norms, she should be entitled to a care person, or, if the elderly living with the family is the norm, to a family member, which would limit the risk of her being robbed. In the instance where she values her freedom more than all the risks linked with being alone, then she should be entitled to an environment in which she can be safe.

The aim of this simple example is to illustrate the limits of the utilitarian approach as "a restrictive approach to taking note of individual advantage in two distinct ways: (1) it ignores freedom and concentrates only on achievements, and (2) it ignores achievements other than those reflected in one of these mental metrics [i.e. pleasure, happiness, or desire]" (Sen 1992, 6). A utilitarian view on human well-being suggests that the goal of development is a state or static condition of a person, that is, a state of satisfaction, and thus it understates the importance of agency and freedom in the development process (Nussbaum 2003). Utility maximization appears to be the ultimate goal of economic agents. They have no influence over the nature of the utility function that they seek to maximize, due to the assumption of fixed preferences. Such an approach leaves no room for adaptation and interdependence, as mentioned before. An individual can pursue happiness in all domains of life, and that can involve the development of capabilities in each domain of life under the constraint of time. As capabilities are developed, this constantly switches the reference point (Kahneman and Tversky 2000) or threshold of happiness (Rayo and Becker 2005). The threshold of happiness is not a lifetime deterministic element of subjective well-being. The argument proposed and developed in this manuscript is that social norms and ideals are a determining factor for life satisfaction and for the development of individual capabilities. This issue needs to be addressed in more detail, since it could also determine the process of adaptation that one goes through to return to this threshold or to change the threshold over time.

Similarly, a major critique against a universal list of capabilities stands on the ground that capabilities are not necessarily best defined to be individual capabilities (Uyan-Semerci 2007). In effect, most potential functionings are achieved through adaptation to one's environment and are therefore socially constructed, or are "social capabilities" (Comim 2008). If capabilities are socially constructed, it seems fair to wonder whether the value individuals attach to different capabilities is determined by common norms of behavior. In that sense, an approach concentrating on entitlement failures, such as the access to a restricted choice set available to individuals due to social norms, could enhance the CA. Entitlement failures represent the notion of failing to have access to commodities and capabilities of one group of individuals relative to other groups of individuals. The interest in using entitlement failures to reduce inequality is mainly regarding the application of the CA. Identifying entitlement failures that prevent one group of individuals sharing a common identity from fulfilling their capabilities seems more straightforward than identifying the capabilities of one individual that he or she wishes to develop. This is because, under the assumption that individuals are social beings, as discussed in the next chapter, individuals' reasons to value capabilities are linked to their social and personal identities.

Chapter 2

Identity, Norms, and Ideals

By linking social systems and individualities, norms of behavior define who we are and how we are perceived, our social identities, and who we are able to become, our human capabilities. Social norms can be broadly defined as tacitly agreed regularities observed amongst groups of individuals. Those rules of behavior are set according to certain standards of behavior, or ideals, attached to a group's sense of identity. Beliefs and ideals draw a picture in individuals' imagination of what a perfect being should be if one is black, white, male, female, American, Chinese, Christian, Muslim, Republican, Democrat, and so on. Unconsciously, this picture sets the benchmark of behavioral expectations. Individuals have multiple social identities and behave according to identity-related ideals, and they also expect others sharing a common identity to behave according to these ideals. Norms of behavior related to these ideals affect people's perception of oneself and others, thus engendering a sense of belonging to particular groups of identity.[1] Individual decisions are therefore constrained by identity-related norms that influence individual "entitlements" or access to resources, both at the level of the household and at the level of the economy as a whole.

Norms can be accepted by a group of individuals although these norms are not necessarily considered acceptable according to other standards of behavior. The interaction between individual behaviors is therefore intrinsically related to what is considered to be acceptable or fair and serve as a basis for individuals' expectations of others' behavior. In the context of the market place, market exchange between different players is influenced by consideration of fairness of what is considered to be an acceptable exchange for a given price. A large body of literature[2] in effect shows how norms and concerns of fairness lead to market failures, including problems of sticky wages and free-rider behavior. However, market exchange often takes place as a result of an accepted norm of fairness according to an ideal behavior. For example, the market value of professional sportsmen is likely to go beyond a socially acceptable price, that is, it can be perceived as unfair by individuals outside this market deal. Instead, the buyer and seller agree on an acceptable price for the expected talent of the player in the new team, which will hopefully make them win all the trophies of the season, or at least meet the team's targets for the season. In this context, the notion of fairness for the market price of a specific combination of human functionings depends on the willingness to pay and to be paid a certain price. However, this willingness relies on the buyer's and seller's perception of a fair exchange according to a common target.

In recognizing that social norms are endogenous to individual decision-making, this chapter does not question the rationality of individual decisions but investigates the benchmark against which the individual decision-maker is rational. Is the individual rational according to a utility-maximization program attached to the consumer identity? Is the individual rational according to a profit-maximization program attached to the producer identity? This chapter makes the proposition that the individual is rational according to a set of optimal behaviors attached to multiple identities that serve a common ideal, as a human being, to function in a living environment, and that social norms act as constraints on the choice set of optimal behaviors. The first section reviews the literature on the origins of social norms in the context of evolutionary game theory and in the light of the empirical evidence on norms and individual decision-making. Then, the second section discusses the links between identity and ideals, which eventually leads in the last section to a discussion on the role of ideals on inequalities of access to human development between identities.

2.1. On the Origin and Stability of Norms[3]

Social norms emerge over time with no explicit historical agreements. The time dimension, therefore, gives an arbitrary nature to norms and makes us wonder how they actually emerge. In modern game theory, evolutionary game theory in particular, social norms are rules defining appropriate and inappropriate behaviors, values, and attitudes that restrict the set of choices available to individuals. In the words of Shoham and Tennenholtz (1997), a social "law" is a restriction on the set of actions available to agents, and it restricts the agents' behavior to one particular strategy. These restrictions are constructed through evolutionary stable strategies in stochastic games, that is, dynamic games composed by the succession of random stages. Then, imitation or adaptive behavior allows a norm to emerge as a stable equilibrium of replicator dynamics. In this framework, social norms bring the time dimension to static games, linking short-run dynamics to long-run equilibria, and serve as a basis for cooperative behavior. Once a stable equilibrium is reached, equity consideration will define whether or not the equilibrium reached remains stable.

In resource allocation, the equilibrium stability is intrinsically related to equity consideration, or at least to what is perceived to be an equitable outcome. In effect, social norms may lead to unequal outcomes in the allocation of resources and yet be perceived as equitable according to a certain ethical standard. Standards of fairness shared by individuals within a group identity provide stability to norms by setting the rules of resource allocation. Once stability is reached, future games are constrained by these standards until conflict with other groups arises and new common standards have to emerge to reach stability. In that respect, Koford and Miller (1991) argue that the equilibrium reached is likely to involve individuals committed to an ethical standard and other individuals not committed to this standard. Then, those committed to an ethical standard have a moral obligation to punish those who violate the norm, which sustains the existing norm over time. The "dominant" standard here will determine whether the outcome is fair or not according to that standard. For example, assuming a shared notion of fairness, Rabin (1993) shows how rational behavior can lead to fair outcomes in a two-person cooperative game. Here, his interest was to model intentions leading to fair outcomes by rewarding kind behavior and punishing unkind behavior. The modeling of intentions is important for the understanding of economic incentives and cooperative behavior. However, if one starts with a shared notion of fairness that accepts unequal outcomes, intentions will necessarily lead to a fair and accepted unequal outcome if inequality is part of the dominant norm of fairness. If the norm is based on an unequal allocation, then fairness as a norm is not necessarily a matter of universal equity but rather a matter of equity according to one specific standard. In effect, it is precisely because there is no agreed notion of fairness between all players that inequality gets perpetuated over time.

The assumption that rational individuals share an identical notion of fairness that makes them agree on a regularity of behavior can be traced back to Hume, one of the early proponents of a definition of social norm. His definition of a social norm is related to a sense of common interest that each man feels, remarks in his fellows, and which carries him along (Hume 1777). The focus on a single characteristic of the individual used to illustrate his point may be worth underlining. In effect, the interest between an individual and his fellows is necessarily common if they all share a sense of identity that is common between them, such as gender in this case. Under the umbrella of norms, modern game theory seeks to theorize the problem of choice restriction. Failures to acknowledge various groups' interests, however, lead to a misunderstanding of inequality in choice restriction leading to inequity in market outcomes. In that sense, by modeling fairness as self-centered inequity aversion, Fehr and Schmidt (1999, 819) provide a convincing attempt to define fairness:

> Inequity aversion means that people resist inequitable outcomes; i.e., they are willing to give up some material payoff to move in the direction of more equitable outcomes. Inequity aversion is self-centered if people do not care per se about inequity that exists among other people but are only interested in the fairness of their own material payoff relative to the payoff of others.

The perspective of self-centered inequity aversion allows various interests to be included in cooperative games, whether it is related to the interest of one individual or whether this one individual's interest is related to a broader group. The sense of belonging of an individual to a group sharing a common interest allows the norm of fairness to emerge between group members, which create a group identity. As developed in the next section, the stratification of group identities makes fairness of the dominant group operate as a dominant rule of behavior in market interactions. Important studies have shown that norms can overtake market mechanisms and become a determining factor in setting market prices. Further work from Fehr and Fischbacher (2002) has illustrated the role of social preferences on market prices, cooperation, material incentives, contracts, and property rights.

In the context of market exchange, unequal outcomes are more likely to be sustained in markets where the exchange is based on the norm of fairness of the dominant group of players. Using various game settings, Fehr and Schmidt (1999) show that fairness plays a smaller role in most markets for goods than in markets for labor. The reason they put forward is that fairness considerations are irrelevant in markets for goods if none of the competing players can punish the monopolist by destroying some of the surplus. The reason put forward here is that the norm of fairness from the monopolist perspective acts as a constraint on the choice set of competing players and restrict all agents' behavior to one particular strategy as a consequence. In the labor market, for instance, groups of interest may be composed of employers, employees, or stakeholders. The norm of fairness may differ between those groups depending on the perspective adopted: whether fairness is understood from the employer's point of view or from the employee's side. In the shirking approach to efficiency wages, employees may deem it fair to shirk on the basis that their wages are lower relative to others' wages. Employers can thus create incentives to increase labor commitment and therefore productivity by raising the pay rate. In line with the gift-exchange relationship, the marginal effort provided by employees responds to signals sent by employers with marginal increases in pay and vice versa. Based on an experimental labor market, Fehr et al. (1998) found that the social norm of reciprocity in a gift-exchange relationship implies that workers will put more effort into their job when they are paid above the market-clearing wage. Other empirical works by Fehr confirm the presence of fairness consideration and its influence on the provision of effort and productivity (Fehr et al. 1998; Fehr et al. 2009).

The market identity of agents influences their perception of fairness and labor market outcomes as a result. A study by Kahneman et al. (1986) found that, in most circumstances, cutting nominal wages downward is socially perceived to be unfair. A cut in wages linked to shifts in demand is socially considered to be unfair and may have repercussions on the firm's reputation. Their research essentially relied on telephonic surveys of American citizens who were asked about the type of price changes "socially" perceived as fair or unfair, that is, what is the overall perception of society to these price changes. However, the market identity of respondents is likely to have influenced their perception of fairness. For example, respondents who are employers may tend to perceive this type of wage cut to be fair cuts according to an objective of profit maximization, while respondents who are employees may tend to perceive this type of wage cuts to be unfair cuts according to an objective of utility maximization.

As part of efficiency wage theories, the fair-wage-effort hypothesis argues that wages are set above market-clearing levels essentially because employers have a perception of how much they should pay for a given work or effort, as much as employees have a perception of how much they should be paid for a given work or effort. In other words, norms attached to fairness affect the remuneration of a certain type of effort. Social norms influence the level of wages through the perceived level of effort associated with the worker's abilities for which the worker has been hired. For employers and employees, both supply of labor and demand for labor depend on the price of labor. However, given that the monitoring of individual performance and effort is difficult, determining the price of labor for a type of worker must rely on a perceived social value of the effort-related capabilities of this type of worker. Assuming that effort is not measurable and varies across individuals, the assessment of the value of a particular type of workers must rely to some extent on the social perception of this type of workers outside the workplace. The norm of fairness influences both employees' and employers' perception of a fair wage for groups of workers sharing an identity such as race, gender, or age. Social norms then place a relative value on the effort of different groups.

In developed economies, increased levels of unemployment for individuals below 25 years, despite high levels of education, and for those above 55 years, despite increased life expectancy, may be in part explained with the existence of fair wages. If within labor markets an ideal employee is socially perceived to be between 25 and 55 years, social groups outside the group representing this ideal will face unequal job opportunities, and this is even more so when markets experience increasing uncertainty. In other words, norms within the society lead to a different perceived value between social groups in labor markets. Stratification economics in effect shows how group disparities in market outcomes can be sustained and exacerbated over time. Stratification economics integrates the significance of group positions and status from sociology with self-interested behavior from economics (Darity 2005; Mason 1995; Williams 1993). Within stratification economics, group positions and status are treated as "produced forms of individual and collective property with both income and wealth generating characteristics and whose supply and demand are responsive to changes in production costs and budget constraints" (Stewart 2008, 803). From this perspective, the existence of gender and ethnic inequalities is the produced outcome of "investments" in social norms that have promoted structured and cumulative advantages for some demographic groups. Norms therefore serve as rules for reproducing advantages for certain social groups at the expense of others. Herd behavior is a powerful factor in explaining the dynamics of group inequalities. Developments in the branch of stratification economics, as described below, show how the interaction of multiple identities influences social interactions and market decisions.

2.2. Identity and Ideal

According to stratification economics in particular, the relative value assigned to social groups is mostly historically determined. In the process of market allocation, this relative value serves as a benchmark or rather as a criterion of optimality to define what job, level of wages, or resources should be allocated to what group. Whether one group is deemed to be worse off or better off in the process of market allocation depends on a benchmark according to which one group is perceived to be worse off or better off. Similarly, the existence of social norms and the recognition of their significant role in market outcomes have questioned the benchmark against which the individual decision-maker is rational. If social norms influence one's rationality by putting constraints on one's choice set, then according to what benchmark is one rational? Ecological rationality[4] suggests that an ecological system emerges out of cultural and biological evolutionary processes, principles of actions, norms, traditions, and morality, and that individuals use "reason...to discover the possible intelligence embodied in the rules, norms, and institutions of our cultural and biological heritage that are created from human interactions but not by deliberate human design" (Smith 2003, 469–470). Norms, rules, and institutions of our cultural and biological heritage are indeed created from human interactions. However, to know whether they are the result of deliberate human design requires a better understanding of human interactions, and more fundamentally it requires a better understanding of the "individual" in social interactions.

Norms and institutions of our cultural and biological heritage influence the benchmark against which one is rational. Since norms and institutions are socially constructed, it makes us wonder about the extent to which rationality should be considered from the perspective of the individual, as an atomistic being, or from the perspective of a group identity whose members share a similar norm of fairness. In a concluding note in his book review on Smith's "Rationality in Economics," Sunder (2009) points out that each person has many "selves," and that the individual self is one of them, but this individual self is "not always the dominant one." Understanding the social being in each individual should highlight how influential are norms in guiding individual decisions and why some identities are more dominant than others according to the context of social interactions. In recent decades, the concept of identity has brought valuable insights to economics by seeking to define what the individual is. The research on identity in effect points in the direction that the individual is a social individual who is rational according to a set of optimal behaviors or identity-related ideals. Identity economics[5] is a relatively new branch in economics since the concept of identity was inserted into the utility function by Akerlof and Kranton (2000). Identity economics seeks to overcome the shortcomings of traditional economic theories in explaining individual behavior, and, in particular, inequalities in market outcomes. The concept of identity is related to the fact that individuals are attached to social categories to which they feel a sense of belonging, such as ethnicity, gender, or age. Insider-outsider models of identity, for example, suggest that individuals are more likely to conform to the insider's norms of behavior and ideal if they feel to be insiders themselves, while outsiders will act differently and create their own sense of identity partly by acting so. In other words, outsiders will conform to the norms of behavior and ideal of the outsiders' group, which will identify themselves as outsiders. Theoretical applications of their insider-outsider model then range from norms of gender and race, to attitudes toward education, and to groups' behavior, such as behavior within the army force.

However, given that the individual is composed of multiple identities or selves, this psychological explanation may generate a view of the individual as a detached social image of the self (Davis 2011). His rationale is that exclusive reliance on social categories to explain social group identity misses out on the fact that social identity is relational, that is, based on roles rather than on categories like insider-outsider models. Davis therefore argues for the possibility of a "personal" identity generated through a self-narrative regarding which type of identity one should conform to according to past circumstances and expectations of future circumstances. An individual is socially embedded as a collection of capabilities that is guided by a personal identity self-narrative of forward-looking and backward-looking self-representations. This view brings an important time dimension to identity, as did the developmental view of ethnic identity mentioned below, but within a broader relational framework of multiple identities. Equally, his insistence on roles rather than categories brings back the importance of social perception in defining the role each identity *should* play in social interactions, which is the perspective adopted in this book.

The cross-disciplinary origins of research on identity in economics question the foundation of traditional economics that adopts the perspective of an economic system starting at the individual level, where the individual is understood to be an unbreakable whole. According to Casey and Dustmann (2010, F-32), the origins of research on identity and ethnicity in particular come from two main approaches in psychology: social identity theory (Tajfel and Turner 1986) and developmental theory (Erikson 1968). Social identity theory focuses on self-esteem issues related to ethnic identity. Developmental theory suggests that ethnic identity varies with age from early adolescence and that acculturation (behaviors, attitudes, and values that may change in different cultures) influences the ethnic identity ultimately achieved through this process of development. Another origin of research on identity comes from the sociological approach of the individual agency, that is, an individual as an agent who actively negotiates his or her roles and relationship with others. On one hand, the sociological approach emphasizes on the two-way relationship between the self and society and how social interactions shape the self. On the other hand, the psychological approach puts a greater emphasis on the identification process "within" the self. For example, in Akerlof and Kranton's (2010) psychological approach to social identities, departing from or conforming to identity-related behavior depends on the degree of anxiety generated by social interactions. As a result, if one departs from his or her insider or outsider norms of behavior, one will suffer from a loss of identity utility. Using identity as a perceptual reference point to go back to is indeed consistent with the fact that social norms affect the level of remembered utility at time $t + 1$ and bias the experienced utility of time t (Kahneman and Krueger 2006) according to this social norm.

The literature on identity brings a totally new perspective to economics redefining the individual as a set of multiple identities constructed through social interactions, including market interactions. The complexity of the individual as a social being means that, to a large extent, perception of identities and associated norms of behavior serves as a basis for an individual's expectations of others' behavior. In the literature, the concept of identity is usually referred to be related to gender, racial, historical, or cultural matters. However, Sen (2005) argues that the concept of identity goes beyond ethnicity, gender, or age, since each individual is composed of a complex set of identities such that each individual experiences a sense of belonging to different groups sharing a common characteristic or interest.[6] Examples of identity could therefore include the following: belonging to a community, caste, religion, or nationality; sharing a common educational background; supporting a sports team; being the fan of a rock star; and so on. Each individual perceives part of his or her identity in the behavior of peers sharing a common identity. In a sense, the social status generated by answering others' expectations of one's behavior contributes to one's sense of "being." People's expectations of others' behavior depend to a large extent on an idealistic vision of the identity with which they interact. Insider-outsider identity models, in particular, have demonstrated the influence of ideals on behavioral norms and, subsequently, on the sense of identity of oneself. For example, in the context of gender identity in Taiwan, Chang (2011) shows that gains in gender identity positively affect self-reported subjective well-being of women while losses of gender identity negatively affect their self-reported subjective well-being. The ideal gender identity here not only serves as a basis for one's expectations of gender behavior but also, by complying with this idealistic gender identity, sustains the social status of each gender.

The idealistic vision of an identity is based on a social perception of perfection attached to this identity, constructed historically, and which can be consciously or unconsciously reproduced over time.[7] As norms emerge, an ideal sets the criteria of optimal behavior, which is identity specific and serves as a basis for social interactions. Belonging to a group sharing a common ideal engenders a sense of identity for its group members. Goette et al. (2006) in effect show how group membership creates social ties that lead group members to enforce a norm of cooperation between them. Being part of a business, country, or international organization creates a sense of identity related to an ideal vision of the business, country, or international organization. For example, professional ethics[8] create a sense of belonging to members of a professional group, which may be through a code of conduct. Members share a common vision of what the ideal professional in that group should be. In a professional context, ethics should in effect guarantee that the professional identity of an individual prevails over all other identities of this individual whenever there is a conflict of interests between the individual's multiple identities. Another example of group identity following an idealistic view of that identity is culture. At the national level, it can be argued that a country's motto creates a sense of belonging to a national cultural identity. To some extent the mottos "In God we trust," "Dieuetmondroit," "Liberté, Egalité, Fraternité" affect behavioral norms and social interactions in the broadly defined American, British, and French cultures, respectively. In the latter culture, for example, the freedom of choice behind the ideal "Liberté" is often constrained by unequal opportunities of choice between groups of individuals, in which case the ideal "Egalité" is not warranted for all. As a result, whenever Liberty and Equality conflict with one another for a group of individuals, belonging to a "Fraternité," or to a specific professional body, may be the triggering factor for strike actions, which can arguably become the norm of cultural behavior.

Institutions create an image of an ideal human being toward which human behavior should tend to. Using the words of the Capability Approach, promoting the capabilities "one has reason to value" (Sen 1999) is influenced by beliefs and ideals that place a weight on the capabilities people deem worth promoting. For instance, the principles of Buddhism mean that little value is placed on material resources and high value is placed on spiritual enlightenment. Equally, the Ten Commandments set the criteria of the optimal human behavior in Judaism and Christianity. In the context of religious, humanist, or philosophical beliefs, these spiritual identities set the criteria of optimality in human development in terms of the benchmark for human flourishing. Ideals serve the benchmark for human development, but also, as a result of optimality point in the allocation, consumption, and production of resources. A critique of traditional economics, for example, is to have created an idealistic view of the market individual who is a rationale homo economicus. Departing from that ideal is perceived to be a matter of imperfection or abnormality, which is sustained by the use of jargon such as "imperfect information," "externalities," "disequilibrium," "market failures," and so on. Similarly, a critique of neuroscience is to have created an idealistic view of the human brain by constructing many classes of average behavior, which means that "if people are identified as individuals only relative to the classes to which they belong, their distinctness as individuals can only be of a typical kind" (Davis 2011, 134). The intention here is not to enter the sensitive debate about the type of ideals worth promoting, but rather to point out that a hierarchy of ideals exists in social and market interactions that set the benchmarks on the norms of cooperation and fairness associated with these interactions.

2.3. Identities and Inequality

In the context of market interactions, ideals of market identities set the criteria of optimality in market behavior: from the production and consumption of commodities to the allocation of jobs and wages in the labor market, and to the allocation of resources within the household. The identity of the profit-maximizing producer or of the utility-maximizing consumer are essential to understand market behaviors, but accepting that market agents have multiple identities lead to a more complex view of market agents. If market agents have multiple identities, it follows that the optimality point in the programs of profit maximization and utility maximization can be influenced by these multiple identities. In the context of labor market interactions, for example, the dominant social identity of the profit-maximizing producer may influence labor demand decisions. Hiring and firing decisions are likely to be influenced by the identity to which the decision-maker belongs, essentially to minimize the uncertainty of dealing with unknown behavioral norms associated with the other groups' identity. Traditional models of labor market explain imperfect competition and discrimination in part as a consequence of asymmetric information. In effect, research has shown that asymmetric information regarding effort between employers (Schönberg 2007), employees (Fehr et al. 2009), or employers and employees (Chang and Wang 1996) may lead to normative judgment and serve as the basis of job and wage offers. The problem of asymmetric information may in effect lead employers to make hiring decisions according to a hierarchy of ideals and associated behavioral norms, which thus impose certain social identities at the expense of others.

In the context of labor market integration of minorities, let us illustrate how employers can be influenced by their own identity as whites, blacks, Hispanic or Asian, immigrant or nonimmigrant, male or female, in the EU and US labor markets. First, looking at European labor markets, the first extensive study on the economic situation of immigrants is by Algan, Dustmann, Glitz, and Manning (2010), who looked at first- and second-generation immigrants in France, Germany, and UK. In terms of employment opportunity, Algan et al. (2010, F25) found that "employment gaps for men in Germany and the UK seem quite similar for first and second-generation immigrants but France has a number of groups in which the second-generation immigrants seem to be doing worse than the first." Digging deeper into that issue, Casey and Dustmann (2010) looked specifically at the own sense of identity of first- and second-generation immigrants in the German labor force to understand its implication on their labor market outcomes, that is, employment, unemployment, and earnings across generations. In a household survey, immigrants were asked about their own sense of identity, whether they felt a sense of belonging to the German identity or to their country of origin. One surprising result concerned second-generation immigrants. They found a positive correlation between the home country identity and employment, and a negative correlation with unemployment. For one standard deviation increase in males' home identity, there was about 6.6 percentage point increase in the employment probability and 2.8 percentage point decrease in the unemployment probability. The main reason put forward by the authors is that the strong sense of belonging to countries of origin may enhance their incentive to draw from ethnicity-based networks to find a job. However, this positive association could primarily be attributed to the fact that second-generation immigrants with a strong sense of belonging to a group of immigrants will have greater employment opportunities in immigrants' type of occupations, thus making use of ethnicity-based networks, than their peers trying to integrate into the "nonimmigrants" group. To enter nonimmigrant types of occupations, immigrants may have to face identity preferences of employers that are similar to the employer's own identities. As a result, ethnicity-based networks represent a means by which immigrants can get access to the German labor market, instead of competing directly with nonimmigrants in jobs where nonimmigrants represent the ideal employee.

Second, looking at the US labor market, unequal access to employment opportunities between identity groups is especially evident in economic downturns. For instance, compared with previous recessions in the postwar era, the recession that started in December 2007 saw a record contraction in US employment, which led many to talk about the "Great Recession." Additionally, since May 2009, the labor participation rate saw a fall of 1.7 percent points, the largest decline since the 1950s (Elsby et al. 2010). Elsby et al. (2010) conducted a thorough analysis of the US labor market in times of recession and specifically over the period 2007–2010. Focusing on the flows in and out of unemployment, they found that the parts of the labor force with the highest unemployment rates and high rates of entry into unemployment comprise the young, less educated labor force, and ethnic minorities. An increasing number of individuals are left out of the labor force or become part of the unemployed labor force. The employed labor force remaining in labor markets is therefore likely to be highly vulnerable to the contextual swings in market confidence. In effect, Elsby et al. (2010) also found that, across the last postwar recessions, variations in employment accounted for between 50 and 80 percent of declines in total hours. This means that the number of workers rather than the number of hours worked by each worker explains the fall in overall employment. The important question raised here is about who is expected to remain in the US labor force during the Great Recession. In other words, how and why existing inequalities are likely to be affected by this fall in overall employment. To investigate this question, one needs to understand the identity preferences of employers in a period of recession—or increasing uncertainty—since, as pointed out by Akerlof and Kranton (2010), identity concerns can influence personal preferences and, consequently, labor market outcomes.

In times of rising uncertainty, identities that are not matching the "ideal" identity from the perspective of the labor demand are likely to suffer most from lack of job and income opportunities. In the US labor force, whites represent the dominant identity of both agents (employees) and principals (employers). In 2008, whites represented 80.66 percent of the full-time labor force and 83.61 percent of the part-time labor force (Current Population Survey 2010). These figures also correspond to the proportion of whites in managerial occupations, since 81.9 percent of US managers are whites, and 62.6 percent are men. A typical manager in the US labor force would therefore be a mature white man, or woman, depending on whether the occupation can be classified as a "male job" or a "female job." As a result of the financial crisis and related recession, full-time employment has fallen significantly across the US labor force, which was accompanied with an increase in the flexible labor force, that is, men and women working part-time (Sahin et al. 2010). This trend is confirmed by the positive growth rate in part-time employment across all ethnic groups. From 2008 to 2009, all ethnic groups have suffered from the loss in full-time employment opportunities. However, the fall in full-time employment opportunities has been uneven across ethnic groups. For example, whites have experienced a fall of 5.98 percent in full-time employment while blacks have experienced a fall of 8.64 percent in full-time employment from 2008 to 2009. On the contrary, part-time workers across most ethnic groups have experienced positive growth rates over the same period. The rise in part-time employment opportunities has also been uneven across ethnic groups. Looking at the highest growth rates across ethnic groups, employers in search of a more flexible workforce seem to have favored whites and Hispanics, with a growth rate in part-time employment of 7.38 percent and 18.66 percent, respectively, from 2008 to 2009, against 4.29 percent for blacks.

Given that the future is uncertain, especially in economic downturns, the hiring or firing decision is very likely to go in favor of the agent of the same social category of the decision-maker in order to sustain this social category over time. The well-known "Glass Ceiling effect" recalls this phenomenon. The Glass Ceiling effect represents the fact that nondomi-nant groups, such as nonwhite and female agents in Western economies, are unable to access higher job positions across the labor market. The dominance of one identity group at the top of high-earning occupations sustains the norm that this elit-ist group is the optimal group in high-earning occupations, thus preventing other groups from breaking the Glass Ceiling. Arestis et al. (2011) have shown that the considerable changes in income distribution caused by the process of financialization over the past 30 years have influenced social norms of fairness regarding the level of wages and job opportunities for men and women across different ethnic groups of the US labor force. Income distribution changes have created and reinforced social norms that have interacted with "fair-wage constraints" to pro-duce and perpetuate "socially acceptable" gender and ethnic identities with different income and wealth characteristics, thus leading to different bargaining powers linked to these iden-tities in the labor market. The formation and persistence of "socially acceptable" gender and ethnic identities have there-fore established a source of intergroup conflict and increasing earning inequalities in managerial and financial occupations.

Two market agents with similar abilities, skills, and education but with two different sets of social identities are likely to face unequal labor market opportunities. The issue of inequality in market interactions between different sets of identities belongs to a wider issue of inequality in social interactions. Multiple identities in effect influence social interactions, including market interactions. The disparities between identities, whether in terms of gender, race, culture, class, caste, sexuality, or age, have not all been paid equal attention in the literature. The most common interest in studies of group disparities is related to gender disparities.[9] One of the reasons for this is empirical in the sense that gender is one of the most visible identities. As a result, a wider range of data is readily available from official institutions to explore gender disparities than for any other group of disparities. However, a main problem behind gender disparities is often found to be that gender is the result of cultural or religious norms of gender roles. Linking together religious, cultural, and gender norms, Ruwanpura (2008) demonstrates how the subtlety of multiple discrimination in the labor market finds its roots in the multiple identities of social interactions. In the UK, a female secondary school teacher was dismissed for wearing a veil in 2006. In the same year, a female checking agent at British Airways was put on probation for wearing a cross on her necklace. Despite sharing a female identity, two different labor market outcomes emerged from these two experiences. Here, the social perception of both gender and religious identities interacted in such a way to produce unequal labor market opportunities for both women.

Inequality between individuals is therefore intrinsically linked to the social identities of each individual. Different combinations of identities can lead to unequal choice sets of opportunities between individuals. More precisely, inequality depends on the way each identity is socially perceived and valued. Individual well-being is consequently affected according to these combinations of identities. The two examples from the literature demonstrate the implications of different interactions of cultural, religious, and gender identities on human development and subjective well-being. First, focusing on cultural and religious identities, the research by Altman and Lamontagne (2004), mentioned in the previous chapter, shows different well-being outcomes between Russian-Polish Jewish immigrants and Southern Italian Catholic immigrants in early twentieth-century New York City. Despite having significantly less purchasing power and more crowded households than Southern-Italian Catholic immigrants, Russian-Polish Jewish immigrants were significantly better-off in terms of human development. As human beings, the two groups experienced different well-being outcomes due to different sets of identities. Assuming that both groups started at birth with similar natural endowments, the social conditioning of individuals in each group led to two different sets of human functionings.

Second, focusing on cultural and gender identities, research by Beutelspacher et al. (2003) found no straightforward positive well-being outcome for Mexican women in Chiapas when adopting contraceptive methods promoted by state family planning programs. Based on six rural communities in Chiapas (Mexico), the adoption of surgical sterilization by women with children was found to worsen their subjective well-being. From the perspective of gender identity, sterilized women felt mutilated and deplored the lack of little children around them as they were used to. From the perspective of cultural identity, it should be noted that most women (65 percent) targeted by this program had already accepted traditional gender roles of housekeeping and child rearing. Sterilized women were forced to adopt the radical contraceptive method they were subjected to, consequently losing the ability to make any choice regarding fertility in the future. Not being able to choose matters related to bodily health on their own also undervalued their ability to think as human beings. Cultural identity therefore mutually reinforced the well-being outcome on gender identity by affecting negatively both gender-identity utility and cultural-identity utility.

The literature on identity and social norms presented here raises a fundamental question on how the individual should be defined. Assuming that each individual in "its" physical body

shape is born with natural endowments, including psychological and physical attributes, these natural endowments at birth
are the results of previous generations' endowments that are
coded in personal genes. Since social norms act as constraints
on human opportunities for development, as shown by the literature, the next fundamental question is therefore regarding the
extent to which these natural endowments are affected by social
conditioning. The line of argument adopted for the rest of the
book, and notably with the approach on exchange-entitlement
mapping, is to think of the individual as a unique combination
of identities whose converging ideal is to function in its environment. Failures to be able to behave toward the achievement
of that ideal are considered to be failures of entitlements to this
optimal functioning.

Chapter 3

Exchange Entitlement Mapping

An entitlement approach called "exchange entitlement mapping," or E-mapping, is developed throughout this chapter and chapter 4 to understand how economic events, such as changes in exchange rate, interest rate, or inflation, affect individual well-being. With his work on poverty and famines, Sen provides a basis for the measurement of exchange entitlement sets faced by an individual in the context of market interactions. E-mapping, in the sense developed by Sen (1981), represents the set of consumption bundles that the individual faces, any of which can be chosen, given his or her endowments. Sen used E-mapping to explain famines, which are due to entitlement failures to food supply, instead of the traditional explanation of a shortage in food supply alone. However, the theoretical framework presented here proposes a dynamic approach to E-mapping augmented with capabilities, from achieved functionings at time t_0 to potential functionings at time t_1 according to the individual's social and economic entitlements. This framework helps to demonstrate that the functionings of individuals become the social entitlements of others, thus leading to the interdependence of functionings between individuals.

As much as individuals have a set of personal capabilities at a point in time and natural endowments from birth, they also have access to limited opportunities inherent to the economic and social system in which they live preventing them from achieving their ideal set of identities. Each individual is a unique combination of identities with a certain hierarchy of related ideals. The converging ideal of all identities is to function in its sustainable living environment. Failures to be able to behave toward that ideal are considered to be failures of entitlements to this optimal functioning. The research objective here is to highlight individual entitlements or, in other words, access to economic and social opportunities of individual development. Individual entitlements are constrained by the interdependence between the choice set of an identity and the choice set of another identity. Thus, by allowing the interaction of capabilities and identities, the aim is to understand the failures leading to the limitation on the development of capabilities over time, whether in terms of economic or social entitlements.[1] Entitlement failures to human development opportunities lead to inequality in the choice set of different identities, which then translates into the impoverishment of capabilities. The concern here is not so much about the conflict of entitlements between individuals, or about the conflict of identities for an individual, which will appear ultimately into a hierarchy of ideals. Rather, the concern here is about the fact that norms set the rules for the interdependence of entitlements and therefore determine this hierarchy of ideals at the personal and interpersonal levels.

Entitlement failures mean that individual freedoms to choose are undermined by the reduced number of alternatives the individual can choose from. In other words, entitlement failures concern the failures of the economic and social environment to provide the same opportunities for all identities to develop the capabilities one chooses to develop. In effect, the environment in which the individual evolves is a crucial element in determining his or her capability set. Restricted access to a socioeconomic environment puts constraints on the possibility to achieve potential functionings, which leads to entitlement failures to achieve these potential functionings. In a sense, it is in the tradition of Keynes who argued that economics should ensure the material preconditions for a happy life, not happiness itself[2] that the approach of exchange-entitlement mapping developed here concerns entitlements, or more specifically entitlement failures, as preconditions that are not available in one's living environment. Individual entitlements to an economic and social system support and sustain the development of individual capabilities. If the economic and social systems fail to sustain the development of individual capabilities, then these entitlement failures should be identified according to the group identity that is affected and be corrected accordingly.

The approach of exchange entitlement mapping (E-mapping) augmented with capabilities puts forward the proposition that social norms are the channels through which the economic environment of individuals affects their opportunities and freedoms to choose different states of well-being and functionings. By including both commodities and achieved functionings in the endowments of the individual E-mapping, this chapter theoretically shows that the capabilities of human beings are built through entitlements, whether economic or social. The exchange entitlement mapping is augmented with capabilities in the first section of the discussion, followed by a discussion of what personal endowments should include in that framework. The third section discusses the complex interactions between identity-related exchange-entitlement mappings to assess opportunities of capability fulfillment between different sets of identities.

3.1. Exchange Entitlements and Capabilities

In its original version, E-mapping is the function that specifies the set of alternative commodity bundles that a person can command respectively for each endowment bundle (Sen 1981, 45). In the discussion here, however, functionings are added to the informational basis of endowments in the individual E-mapping, as shown in the next section. E-mapping becomes the function that specifies the set of alternative commodity bundles and potential functionings that a person is entitled to, or has access to, respectively, for each personal endowment of commodities and achieved functionings. The issue of endowment will be addressed further in the next section, but a more detailed definition of functionings is required at this stage. Functioning is defined as an achievement of a person, what he or she can do or be according to the commodities available and functionings achievable in his or her living environment, while a capability represents the ability to achieve. More precisely, an achieved functioning, as opposed to potential functioning, is the conversion of an available commodity into a functioning, depending on personal conversion factors—that is, metabolism, physical condition, sex, intelligence (Robeyns 2005b, 99). As pointed out by Anand et al. (2005), capability and functioning may be difficult to distinguish due to the circular nature of the relationship between the two (Anand et al. 2005, 43):

> Is health which limits your activities a capability in that it restricts the potential choices you can make, or is it rather a functioning, the result of the choices you made from your capability set to, e.g., smoke or drink?...Perhaps the answer lies in Nussbaum's point that what people choose to do should not be the focus of policy makers but rather that enhancing the choice set available to everyone (even smokers and the obese) should be.

The policy focus on choice set's enhancement made in the above statement is in line with the idea of entitlement failures, whereby removing sources of entitlement failures to opportunities would enhance the choice set available. Looking at the example above from the identity perspective instead of the individual perspective, the matter of choice restrictions raises two questions. First, what is the proportion of choice linked to free individual choice? And second, what is the proportion of choice linked to identity norms of behavior in place in the living environment of the individual and followed by the individual? The connection is that individual freedom of choice between different opportunities is constrained by behavioral norms attached to the individual's set of identities. In this sense, absolute freedom is not possible, and mutual interdependence amongst individuals implies relative freedom. The extension of some individual freedom may impede on others' freedom. Freedom therefore depends both on whether there is something to choose from and whether or not the choice is blocked by others (Veenhoven 2000). The freedom to develop capabilities depends extensively on social norms in place in the living environment of the individual.

From the E-mapping perspective, while the definition of exchange-entitlement per se illustrates the opportunities aspect of freedom, the external macroeconomic shocks affecting the exchange entitlement mapping, say from one period to another, depict the process aspect of freedom, that is, the processes by which opportunities can be transformed into achieved functionings. One could have many opportunities but only be entitled to a restricted set of choices, because of his or her social identity and its ranking relative to others. In effect, one could consider thinking about an ideal set of choices, but one would remain confronted with the reality of the choices actually available within one's living environment. To a large extent, actual access to, or entitlement to, a choice set for an individual is constrained by the actual access to this choice set by others. The importance of the meaning of entitlement for the E-mapping framework makes it worth pausing for a moment on its definition. Looking at the New Palgrave Dictionary of Economics, Steiner argues:

> In the strong sense, an entitlement is something owed by one set of persons to another…[and that] a weaker form of entitlement may be said to pertain to those of a person's activities which, while not specifically protected by obligations in others not to interfere, are nevertheless indirectly and extensively protected by their other forbearance obligations…So, broadly speaking, persons' strong entitlements may be construed as conjunctively constituting their spheres of ownership, while weak entitlements and their unprotected liberties constitute the fields of activity within which they exercise the powers and privileges of ownership. (Steiner 1987, 149)

The entitlements considered as part of the E-mapping here refer to a weak sense of entitlements where the fields of activity are affected by power relationships between groups of individuals. These relationships of power can lead to an entitlement failure of the socioeconomic environment to guarantee the privileges of ownership, which themselves stand on legal or moral grounds.

> Entitlements may be either legal or moral. Sets of legal entitlements tend to reflect the multifarious demands of various customs, moral principles, judicial decisions and state policy. A set of moral entitlements, on the other hand, is commonly

derived from some basic principle embedded in a moral code involved... In many single-value codes (such as utilitarianism), entitlements are instrumental in character. (Steiner 1987, 149)

Legal and moral entitlements are part of the socioeconomic environment in which the relationships of power act. This means that the legal and moral codes can lead to failures of the socioeconomic environment to provide access to potential functionings, or, in other words, to failures of entitlement to the goods and services that could be transformed into achieved functionings. Legal and moral systems generate norms of behavior that affect access to goods and services and subsequently to potential functionings. Contrasting with Steiner's view, the approach adopted here makes use of two types of entitlement, namely, economic and social, and here legal and moral entitlements belong to the generic consequences of social interactions embedded in social entitlement, which include the rules of entitlement, that is, social norms.

Economic entitlement represents actual access to bundles of commodities and potential functionings provided by a given society, either through private, or public, or international channels. Actual access also includes gifts and public goods, that is, entitlements that are related to identity ownership. For example, access to health services is dependent upon the country provision in which one lives, either through public or private health care, and this access is therefore related to a national identity. Therefore, economic entitlement represents actual access to bundles of commodities and potential functionings provided by a given society either through private, or public, or international channels, and that does not necessarily constitute a legal right *per se*. In effect, economic entitlement may arise from illegality, such as the purchase of illegal commodities. Examples of economic entitlement such as access to prostitutes for the client or access to cocaine for the drug addict—where a person pays someone who is willing to supply what that person demands— represent economic entitlements that are, in many countries, not supported by law. Similarly, social entitlement represents actual access to a social or political system with specific norms and values generated historically. For instance, a divorced parent may be legally entitled to meet his children, but in the case where means of transportation are not provided by the market economy in which this parent lives, it will prevent him from meeting the children. If valued by the parent, the functioning of parenting in the sense of affectionate support is therefore affected. In this example, a legal entitlement cannot be realized without the support of economic entitlements. Therefore, economic and social entitlements appear to be two separate concepts, the former being the result of market interactions only, and the later being the result of social interactions. Indeed, capabilities such as the ability to work, the ability to earn an income, the ability to sustain family needs, and the ability to be economically independent are diverse capabilities related to economic entitlements, that is, the external economic environment, such as job opportunities, equal pay, inflation level, and so on.

Social entitlements are the social and environmental conversion factors of commodities into functionings (Robeyns 2005b, 99) specific to the environment in which the individual lives. These include geographical conditions, social norms, public policies, legal frameworks, religious values, power relations, and human and legal rights such as protection from abuse, violence, and discrimination, among other examples. Social entitlement is considered to be endogenous to the individual's living environment, in the sense that, over time, individual achieved functionings have repercussions on the social entitlements of others. The society is composed of individuals with different sets of identities whose actions, driven by perceived ideals, lead to an evolution of social entitlements. Consequently, entitlement for different identities can be distinguished from one another depending on the nature of the individual's environment in which relationships of power between identities are exerted. The market power of firms defines the economic entitlements of their workers, that is, the real wage and benefit entitlements of workers. From the point of view of the firm, the wage should be equal to the marginal revenue product. From the point of view of the worker, the wage should be his or her opportunity cost. If a female worker's opportunity cost for an hour of household work in terms of labor market work is socially considered to be lower than a male worker's opportunity cost for the same hour, then discriminatory practices in the labor market are likely to occur and her wage will be lower than his wage for the same job position. Based on a moral code of socially perceived fairness, the social status of individuals, or gender identity in this case, defines economic entitlements of male and female workers.

In Sen's writings, on which the capability approach used here is based (*Poverty and Famines*, 1981), two points need to be made in order to underline how the concept of entitlement relates to identity. First, the concept of entitlement relates to rules of legitimacy between sets of ownerships whereby "an entitlement relation applied to ownership connects one set of ownerships to another through certain rule of legitimacy. It is a recursive relation and the process of connecting can be repeated" (Sen 1981, 1). Since his framework applies to market interactions, the rule of legitimacy depends on trade-based entitlement, production-based entitlement, own-labor entitlement, and inheritance and transfer entitlement. In the context of social interactions, however, the rule of legitimacy is based on identity-based norms of behavior and relations of power between identities. Second, in Sen's case, entitlement is also used in the sense of "access to," and especially in the context of famines, where he first used E-mapping to look at access to food supply. Understanding entitlement in the sense of "access" means that the concept of entitlement does not necessarily require a prerequisite of human rights or "fundamental entitlements" as Nussbaum would argue (Nussbaum 2011). For instance, over the year 2007, the world prices of both corn and wheat rose dramatically, which could have been due to the increase in the demand for the grain-based fuel, ethanol, or due to bad crops and severe droughts, or due simply to market speculations. A basic ingredient of Mexican diet is corn to make tortillas. A basic ingredient of the Italian diet is wheat to make pasta. As a result of the increase in international prices of wheat and corn, pasta and tortillas became more expensive.

Therefore, the economic entitlement to higher-priced traditional ingredients such as corn and wheat may have changed the social entitlements to cultural food and traditions in both countries. A possible outcome is that, provided that prices kept on rising, both cultures may have had to substitute cheaper food for their traditional ingredient, thus changing their cultural identity. The point here is that the two-way relationship between economic and social entitlements, or "access to," exists without the necessity of the concept of human rights (although Italians could claim, in a dramatic way, that access to pasta is a human right).

Capability well-being relates to the development of personal capabilities, given the commodities and opportunities available in the environment in which the individual evolves. Well-being in the CA is described as an index of the person's functionings from which he or she derives positive or negative feelings. In his Tanner Lectures, Sen (1987) argued for the complementarities of self-evaluation and standard evaluation when talking about measuring living standards. While standard evaluation places a person's living conditions in a general ranking in terms of some social standards, self-evaluation is how the person would judge his or her standard of living *vis-à-vis* the positions of others (Sen 1987, 30). In this context, up to a minimal point to fulfill basic capabilities,[3] living standards strongly influence individuals' welfare. One of the debates on the CA is about defining what basic capabilities are. The main question is whether or not a threshold should be set, below which basic capabilities should be supplied by international, national, and nongovernmental institutions.[4]

Well-being in the CA is seen as the value that a person gives to the functionings achieved, while an advantage represents the opportunities a person has, of which only one will be chosen. The value of the functioning is a personal judgment coming from self-evaluation, which essentially relies on identity-related ideals. Personal judgment is socially influenced and also depends on personal characteristics (physical and psychological attributes) and personal identities. In the CA, however, it has been argued that this value is derived from the use made of a commodity, in the sense that utility is derived from the individual capability to function. Given that Sen's approach focuses on market interactions only, the causal link between commodity and utility depends on the individual capability to transform a commodity into utility (Sen 1999). Thus, the causal link according to the CA (Clark 2006) would be

Commodity → Capability (to function) → Function(ing) → Utility,

instead of the Utilitarian approach arguing for the following causal link between commodity and utility:

Commodity → Utility

From the CA perspective, the necessary and sufficient link between commodity, capability, and utility means that capabilities are necessarily derived from commodity use. A commodity is transformed, according to a personal capability to function, into a functioning from which utility can be "extracted." In Sen's words, a commodity is defined according to its characteristics (Sen 1985), in the sense that it depends on the individual capability to extract his or her own functioning. Commodities are mediated through personal conversion factors, which then generate their impact on subjective well-being. Therefore, a commodity is no longer a source of value *per se*; rather, the opportunity to enjoy a commodity is a source of value.

Outside the context of market interactions, however, and defining the individual as a unique combination of identities, functionings can exist without the prerequisite of a commodity, since they can result from others' functionings. Being able to kill, to love, or to insult someone else does not result from some characteristic of a commodity; neither is the possible utility coming from these capabilities the result of commodity consumption. Rather, it seems that some capabilities that people enjoy are the result of the interaction between individuals without the intermediary of a commodity. Including both commodities and functionings in the E-mapping will allow

individuals, as a set of identities, to be entitled to both potential functionings and commodities where commodities are no
longer the sole source of subjective well-being or capability fulfillment. Therefore, in the context of social interactions, the
proposition adopted for identity E-mapping is of a circular relationship as follows:

Identity → Capability (to function) → Function(ing) →
Utility→ Identity

and where market interactions are a special case of social interactions, it is as follows:

Identity → Commodity → Capability (to function)→
Function(ing) → Utility → Identity

Adopting such a perspective allows us to argue that inequality in the impact of economic events across different identity-
related groups of individuals can be understood in terms of
entitlement failures to capabilities that lead to these inequalities. These inequalities are then reproduced if the norm of fairness setting the hierarchy of identities does not change over
time. For policy evaluation, this analytical perspective allows
to take into account the social impact of economic events and
policies on diverse social identities. One of the strengths of
the CA is to consider capabilities to be human resources that
cannot be all market-traded: some can certainly be bought or
sold, while others can be taught or learned, and received or
given. This includes capabilities related to confidence, knowledge, and sexuality, or any capability related to Nussbaum's
list of capabilities, mentioned in the previous chapter, which
are valuable for well-being evaluation. Including achieved and
potential functionings in the original E-mapping allows us
to go beyond the "economic" well-being referring to income
and work opportunities, and to go beyond the commoditization view of the household by assessing the effects of possible
changes in all domains of life, whether they are social, cultural,
familial, individual, that is, identity-related. Identities can be

social, economic, political, religious, geographical, historical, and so on. In that sense, resources are not solely material.

3.2. Endowments: Commodities and Achieved Functionings

Social and economic entitlements vary from one country to another, which makes each market economy unique in the way in which it performs. From the labor force to goods or services, it seems that any commodity and most human abilities can be potentially bought, sold, and produced. Apart from the ethical limits on what can be bought or sold (slavery is an example), there are human capabilities that lose their meaning if sold, such as trust, love, and commitment. For instance, if a commitment were to be sold, then it is no longer a commitment, since someone else could come along and pay a higher price than the former buyer. Each individual has different levels of skills and functionings, which therefore constitute a heterogeneous labor force. Each individual i has an endowment structure at a point in time which is represented by x, with s as a subset of x, and which can be defined as a vector x_i with n-achieved functionings and commodities at time t. Endowments are assumed to be composed of both commodities already consumed and achieved functionings, dependent upon the personal conversion factors of the individual, as already mentioned above, including metabolism, physical condition, sex, and intelligence (Robeyns 2005b, 99). "Commodity" refers to any good or service that can be bought, sold, or given, as well as public goods, that is, goods and services that are nonrivalrous and nonexcludable, provided both by market interactions and by governments. All other items that individuals are able to give, such as words and gestures, are considered to be achieved functionings. Therefore, the informational basis of endowments and entitlements includes commodities and achieved functionings for the former, and commodities and potential functionings for the latter. Endowments constitute the items that the individual owns and/or transforms and that can be given or exchanged, that is, any ownership or useful item that enables the creation and/or sustainability of personal and others' functionings. In that sense, most capabilities can be described as social capabilities (Comim 2008), since they depend on the opportunities faced by individuals and the processes by which these opportunities can be affected.

The notions of gift and exchange, for money or for other resources, are important for this approach on E-mapping, since the line between what can be bought, sold, or given is quite thin, whether it is a commodity or a functioning. Indeed, working may be one example of a functioning that can be sold. Similarly, public goods such as education or health[5] can be entered into the individual exchange-entitlement mapping as functionings: being literate or illiterate and being treated and healed or not healed. In that instance, a functioning would be "being literate" or "being illiterate" and a capability would be "the ability to be literate" or "the ability to be illiterate." Another example would be of someone using his or her ability to make violent gestures toward other household members, which, whether it is legally or socially sanctioned, becomes a social entitlement for the victim. The extreme case of honor killings would be a social entitlement failure for the victim to be able to live in a secure environment, resulting from the offender's perception of certain norms of behavior. Hence, entitlements do not necessarily require having a positive impact on individual well-being and are thus considered to be entitlement failures if negative—negative in the sense that it restrains the individual capability to function. Capabilities depend on individual endowments—personal character and genes, or personal conversion factors, as Robeyns (2005b) calls them—and also on economic and social entitlements—or social and environmental conversion factors.

Endowments are composed of any commodity and achieved functioning as what the individual recognizes to own and to make use of. Indeed, using this definition of endowments allows us to go beyond the view of pure commoditization of the household, thus including functionings that are the result of household members' and social interaction. Also, it allows us to not ignore the importance of public goods, that is, those that are nonexcludable in provision, nonrivalrous in consumption, and which provide and sustain various capabilities, such as being educated or healed through the education and health systems, respectively, when those are, in effect, public goods. Individual E-mapping starts at the household level, since the household represents an essential place for human development opportunities, together with the local community, neighborhood, and extended family, where children are taught to function according to certain norms of behavior. The household is, however, the first environmental unit where an individual learns about his or her identities, such as being the son or daughter of X, being the brother or sister of Y, or being the grandchild of Z. From childhood, an individual is thus entitled to, or has access to, commodities and potential functionings according to personal conversion factors of commodities into functionings, but these are also dictated by social entitlements, that is, social ties specific to the environment in which the individual lives. Social norms are part of the individual's social entitlements in the sense that social norms vary according to the perceived set of identities of the individual, which is dependent upon the society in which the individual lives. For instance, gender norms vary according to whether one lives in Scandinavian countries or Latin American countries.

Individual endowments represent the set of achieved functionings and commodities at time t, including goods, services, and public goods that each person happens to own or have use of. At the household level, a part s of the individual set is shared with the household set depending on individual choice, and therefore s becomes part of household endowments. Individual choice is assumed to be influenced by social entitlements, that is, by the social perception of identity-related optimal behaviors shaped by institutions. Individual choice is made on the basis of identity utility under bounded rationality, which may be based on altruistic behavior, individualistic behavior, or both. Household endowments are represented in the form of a matrix. The reason for assuming household endowment to be a matrix rather than a sum of individual endowments is that not only is part s of x the set allocated to household endowments, but also because this part s can be changed at any time given individual and extended E-mapping. Consequently, the matrix X (Sen 1981) describing the endowments of household k can be written as follows:

$$X_k = \begin{pmatrix} s_1 i_1 & \cdots & s_1 i_m \\ \vdots & \ddots & \vdots \\ s_n i_1 & \cdots & s_n i_m \end{pmatrix},$$

augmented with n-achieved functionings and commodities $s \subseteq x = 1,...,n$; and m-individuals in the household, $i = 1,....,m$. According to the living environment of an individual, this means that individual endowments are affected by the household endowments both in terms of commodities and functionings. For instance, a favorable environment for a child to grow and acquire a good education gives him the chance to develop skills and confidence that will widen his scope of market opportunities. In that case, the household E-mapping is the major contributor to future individual endowments. Similarly, individual functionings and commodities contribute to household endowments, since other members may also benefit from the access to these functionings and commodities. If one's capability is the ability to work, let us say in a factory, a weekly income is brought into household endowments. Assuming that currency devaluation hits the economy and nominal wages are kept constant, the resulting real wage received by the industrial worker is therefore lower than before the devaluation. The income thus brought into household endowments is lower than before and it restrains the purchasing power that the household previously enjoyed. A possible consequence could be that previous nonworking household members now have to enter the labor market in order to maintain the level of income necessary to sustain, if possible, the same level of consumption. Here, relative income *vis-à-vis* other members is also an issue and depends to a large extent on the norm of relative income between household members, that is, whether all members are entitled to seek income in the labor market, or whether all members sharing an identity are entitled to seek income in the labor market.

According to Sen (1981, 45), E-mapping is the function that specifies the set of alternative commodity bundles that the person can command, respectively, for each endowment bundle. Including functionings in individual and household endowments allows us to go beyond the commoditization of the household. Lewis and Giullari (2005) in effect argued for the impossibility of fully commoditizing care, as subjective factors are also part of care responsibilities: "care is embedded in personal relationships of love and obligation, and in the process of identity formation" (86). The argument against commoditization, also sometimes defined as commodification, refers to the fact that certain aspects of the household life cannot be materially accounted for. Love, care, and commitment are a few examples of individual functionings that cannot be sold on a marketplace without affecting their meaning. Similarly, personal relationships in social interactions would be perceived differently by the recipient from an activity provided through the market, even if to some extent the activity were the same—examples include care of the elderly, child care, or sexuality. The possibilities offered by the CA might overcome such problems of commoditization and nonmaterialism by evaluating individuals' endowment of, and entitlement to, both commodities and capabilities. Thus, entitlement failures are not only failures of access to commodities but also failures to achieve potential functionings. Concerning the labor market, access to employment and access to real wages are two economic entitlements that can be transformed into functionings through the personal conversion factors of commodities into functionings. Concerning the household, the same culture can bring different social entitlements for two individuals within the same household. Indeed, differences in social entitlements can be according to the gender or age of each individual within the household. Conflicts of identities such as gender or age are reflected in the hierarchy of identities emerging in social entitlements, which are sustained through a socially acceptable norm of fairness attached to that hierarchy. For example, using the collective model of the household, Dauphin et al. (2011) have shown that both males and females between 16 and 21, and daughters irrespective of their age, have become decision-makers within British households.

Cross-country comparison would be needed to know whether this is a universal norm, but, from a historical perspective, this was certainly not the norm in previous generations.

3.3. Identity E-mapping

The literature on individual well-being, whether in terms of subjective well-being or capability fulfillment, tends to suggest that individual well-being exists in accordance with an ideal, which set the criteria for individual behavior. Utility maximization is perceived to be according to an optimal equilibrium point, which recalls the set point of Easterlin (2005) where individual happiness tends to go back to, or the reference point in the prospect theory of Kahneman and Tversky (1979). Similarly, capability fulfillment is perceived to be either according to the value individuals attach to their capabilities (Sen 1999)—this value depending on individual beliefs—or according to an "idealistic" list of fundamental entitlements as put forward by Nussbaum (2000; 2003; 2006; 2011). Nussbaum uses the term "entitlement" in the context of a set of fundamental entitlements related to her list of capabilities. However, it seems that her meaning of "entitlement" refers more to inalienable human rights than to a problem of physical access to commodities and capabilities. Equally, she believes that some individual freedoms are more important or central than others, in the sense that institutions should provide basic human capabilities. Her approach on capabilities is that an individual is entitled to fundamental capabilities that are given by institutions, which could be national, international, or nongovernmental. To cite a few examples, her list of entitlements includes the entitlement to a healthy life, the entitlement against violation of bodily integrity, and the entitlement to pleasurable experiences. As she put clearly in Nussbaum (2011), her list of entitlements reflects one view of what ideal entitlements should be. There are, however, as many ideals as there are identities, and a socially acceptable hierarchy of ideals leads to entitlement failures to pursue a personal hierarchy of ideals. In a sense, offering an idealistic list of entitlements helps to highlight entitlement failures to human rights, or, in other words, to assess the discrepancies between these ideal entitlements and actual entitlements of an individual's living environment. However, insisting on the separateness of individual capabilities misses out on the fact that inequality in entitlements between individuals is linked to identity, as discussed in the previous chapter.

Considering the individual to be a unique combination of identities whose converging ideal is to function in its environment, failures to be able to behave toward that ideal are considered to be failures of entitlements to this optimal functioning. The ideal human being one wishes to become changes over time, which reflects changes in identities. As a teenager, one may wish to become a rock star, while, ten years later, this same individual may wish to pursue ideals attached to a spouse, a soldier, a football player, or all three combined. Social identities that are part of an individual's endowments such as race or gender are, however, constant over time. Assuming that there is one E-mapping per identity, each identity is entitled to a set of social and environmental factors attached to this identity, enabling it to turn potential functionings into achieved functionings. In market interactions, the extent to which potential functionings can be transformed into achieved functionings depends on access to commodities related to this identity's entitlements. For instance, let us assume there is one E-mapping for being white, one E-mapping for being African American, and one E-mapping for living in Washington, DC. The American Human Development Project (2010) of the Social Science Research Council shows that whites in Washington, DC, live, on average, a dozen years longer than African Americans in the same city. Combining these three E-mappings in different ways leads to different outcomes of human development. In social interactions, the extent to which potential functionings specific to an identity can be transformed into achieved functionings depends on this identity's entitlements. Potential functionings specific to an identity will be transformed into achieved functionings specific to the unique combination of identities that the individual is composed of. Now, taking an individual with a unique set of identities—of being a man with American citizenship, father of two children, son of a 100-year-old mother, and a football fan—will have a different prospect of human development if he is African American in Washington DC, than if he is white in the same city. Entitlement failure here is represented by the impossibility to transform identity-specific potential functionings—that is, living in Washington DC—into individual-specific functionings by missing out a dozen years of life. It is an entitlement failure that is identity-specific as an African American and as a human being who cannot function at full capacity. The point here is that by recognizing that each individual is a unique set of identities with distinct capabilities, entitlement failures of a social identity to function in his or her living environment are a source of inequality in human development.

In the present approach, E-mapping is the function that specifies the set of alternative potential functionings and commodities that the person has access to, given what he or she already owns as achieved functionings and commodities. The terms of ownership depends on the individual perception of property, whether it is perceived as personal without being collective or collective at the level of the household or beyond and therefore perceived as being personal (public goods are one example). A car can be perceived as a collective good within a household while a bicycle might be perceived as a more personal good, although it may be shared in use. In the context of identity E-mapping, the concern is to find out the potential functionings and commodities an identity expects to have access to in the surrounding living environment and given his or her personal endowments of achieved functionings and commodities. The terms of entitlements depend on a social perception of identity, whether the identity is perceived to be entitled to or not entitled to commodities and potential functionings. At time t_0, an individual with a unique set of identities and related achieved functionings is entitled to a restricted set of choices that is specific to the set of identities to which the individual belongs. Potential functionings at time t_1 are determined first by the choices made by others over each identity's choice set at time t_0, that is, social entitlements, and second by the economic conditions affecting this choice set between time t_0 and time t_1, that is, economic entitlements. First, if each identity is socially perceived to be entitled to certain commodities and capabilities, then each identity is constrained to choose within this choice set. If one is entitled to another identity's choice set, choosing within this choice set makes this identity become part of the unique combination of identities of the individual. Second, economic conditions can widen or restrain an identity's choice set depending on the social perception of this identity relative to others. As a result, the potential functionings at time t_1 transformed into achieved functionings become part of the social entitlements of others at time t_1. In other words, the functionings of individuals become the social entitlements of others since there is always one identity in common, that of being a social human being. Thus, the social perception of identity entitlements leads to the interdependence of functionings between individuals.

Let us illustrate identity E-mapping with examples within the smallest unit of individual functionings, which is the household. Individual E-mapping is strongly influenced by household endowments. In effect, parents tend to develop children's capabilities, given the parents' own access to commodities and potential functionings. Before education became the norm in Western countries, it was not uncommon for job skills to be passed on from one generation to another, such as in the case of merchants or artisans. This was due to limited opportunities for the development of other capabilities, which also rendered the ascension of the society ladder very limited. Again, social and economic entitlements are crucial in influencing the development of capabilities, that is, individuals are the product of their environment. For instance, household child care may bring other capabilities that paid child care does not, and *vice-versa*. The level of nutrition, and thus children's ability to survive, relies on access to food and the level of income as much as the ability of self-esteem and self-confidence rely on parents' ability to provide them with affective support. Family care significantly differs in its outcome on child welfare from care undertaken for monetary payments outside the home. Evidence is mixed in this area. For example, research suggests that stress levels rise among young children when they enter child care (Ahnert et al. 2004). In this research carried out in Berlin, researchers tracked the reaction of 70 toddlers to the separation from parents and homes. Others found that paid child care meant higher IQ and better ability to speak for children in child care than for children who were not, regardless of the status of their parents (Brooks-Gunn et al. 2003). What these results suggest, however, is that children develop different capabilities according to the environment in which they evolve.

The household endowment matrix allows us to assess individual access to resources by identity, such as age, sex, or market position. In other words, the question is how much of individual entitlements can be explained in terms of identity entitlements. The allocation of resources within the household is therefore a determinant factor of individual E-mapping, which leads us to wonder who decides the allocation of resources. Early studies on household behavior relied essentially on a unitary model of the household, a view according to which a single decision-maker is the head of a small factory producing output with the help of raw material and labor power (Becker 1976). Models have moved a long way since, notably in the context of intra-household conflicts and bargaining theories in household economics.[6] Bargaining theories, in effect, stress the importance of the different objectives brought by each individual in a bargaining process to allocate resources within the household. The collective model of the household essentially puts forward the idea that the valuation of household utility by other household members is also a determining factor of one's sense of utility. Other household members' valuation, however, relies on imperfect information, which means that recourse to social norms is often the only way to enforce a cooperative solution (Katz 1997). In this sense, Sen (1990) argues in favor of a cooperative conflict, a situation in which two decision-makers need to cooperate over the use and allocation of resources among household members. Since resources are scarce and individual preferences are heterogeneous within the household, this can involve some conflicts.

In the collective model of the household, the share of individual earnings influences the balance of power between decision-makers. In effect, evidence suggests that, to maximize household utility, bargaining power should be evenly spread between the spouses rather than be left to a dominant member, as argued by the unitary model (Lancaster et al. 2006). At the household level, it is therefore likely that household utility will be linked to norms of behavior attached not only to a cultural identity but also to each individual's social identity, including gender. The maximization of household utility is based upon a concept of optimality, that is, a single optimal allocation of resources. However, if identity influences resource allocation, then is it possible that identity also influences optimality? From the perspective of gender identity, modeling the collective approach to the household requires a serious account of both male and female bargaining powers (Basu 2006, 563):

> If we are to take the collective approach...seriously and want to model it as a game, we face a serious problem: who are the players? Note that in the collective approach the agents are the man and the woman but the decision is taken by a mythical hybrid that is a weighted average of the man and the woman.

The mythical hybrid in the collective approach is therefore a single decision-maker with both male and female identities. The bargaining power of each identity then serves as a basis to know which identity is more dominant in the idealistic role of mythical hybrid. However, if gender identity is influenced by cultural norms on gender roles, then this mythical hybrid may be the result of a social perception of the household optimal resource allocation. The literature has shown evidence that social norms influence intra-household resource allocation,[7] household headship,[8] and household food security and child nutrition.[9] Norms bind the interdependence of people's choice sets, which is especially evident within the household since both preferences and opportunity sets are interdependent.[10] The current empirical literature shows that social norms act as rules of entitlement between identities in the allocation of resources at the household level. The household survey presented in chapter 5 shows the extent to which the benchmark of optimality in the household maximization program is based on gender and cultural identity. Ideals attached to gender and cultural norms determine identity entitlements and therefore influence both subjective well-being and access to capabilities, since they set the benchmarks for behavioral norms, within and outside the household.

3.4. Conclusion

To a large extent, inequality of opportunity sets between identities is related to identity entitlements or, in other words, to the access to resources and capabilities deemed socially acceptable according to one's social identities. Narrowing the individual E-mapping to identity E-mapping within and outside the household allows us to understand that market and social interactions are ruled by a perceived optimality in the consumption, production, and allocation of resources. Poverty and capability deprivation emerge out of inequality of access to resources and capabilities, which affects the well-being of individuals as consumers, wage earners, household members, or simply as human beings with a unique combination of identities. Poverty is the result of entitlement failures to food supply, job, or income, and it affects an individual's opportunities to achieve human functionings. Poverty is therefore the result of entitlement failures in the sense of failures of the social and economic environment to provide capabilities, such as employment opportunities and access to commodities, goods, and services, allowing the sustainability and/or improvements of human functionings. As a unique set of identities and related functionings, each individual has a unique perception of reality. Ideally, this perception should find a common ground in its human identity with an optimality point of functioning in a sustainable living environment.

Chapter 4

External Shocks on E-Mapping

Transforming a potential functioning into an achieved functioning depends extensively on the freedom to choose between different potential functionings, given the set of opportunities offered by the economic, political, and social environments of the individual. Basu (1987) rightly raised the question of whether individuals were really free to choose any set within their opportunity sets. His answer is negative since the actual choice of a person is dependent upon the actual choice made by others. The interdependence of individual choices means that some resources chosen by one individual at a given time cannot be chosen by another individual. The interdependence of individual opportunity sets has also been identified by several authors as being linked to interdependent preferences between individuals, especially in the context of group identity, as will be discussed below in the context of external shocks on identity E-mapping.[1] The solution to the problem of interdependence proposed by Basu is to find the definition of opportunity, which accounts for "the interdependence between one person's choice and another's opportunity set and since a person's choice depends on his opportunity set, ultimately we shall have to deal with the interdependence of opportunity sets" (1987, 75). In this chapter, the use of entitlements by identity embraces the issue of interdependent opportunity sets by looking at the impact of external shocks on the economic entitlements of identity E-mappings. At one point in time, the impact of these external shocks on identity E-mapping is dependent upon the social norms that set the rules of interdependence between individual opportunity sets.

Including both commodities and achieved functionings in individual endowments, the E-mapping framework argues that the achieved functionings of an individual become part of the social entitlements of others, thus leading to the interdependence of opportunity sets. Due to the interdependence of opportunity sets, potential functionings are determined by the interaction between social entitlements and economic entitlements. The purpose of this chapter is to investigate the way economic conditions affect the choice sets of an individual, which is considered to be a unique combination of identities. In other words, the external shocks of economic events and policies affect the economic entitlements of each individual by affecting one or more identity E-mappings. In the first section of this chapter, the first step of investigation is to look at the interdependence of entitlements in order to grasp the mechanism involved when an external shock affects this interdependence. By making identity choices, individuals will affect the potential functionings or choice set of other identities. If each identity is socially perceived to be entitled to certain commodities and capabilities, then each identity is constrained to choose within this choice set. Economic conditions can widen or restrain an identity's choice set, depending on the social perception of this identity relative to others.

Following the first section, the evidence supporting the E-mapping approach to inequality using the *maquiladora* worker's identity starts. The *maquiladora* industry can be described as wholly foreign owned or Mexican owned subsidiary plants, originally on the Mexican border, for the assembly, processing, and finishing of duty-free foreign materials and components into products for export, essentially to the United States. The *maquiladora* industry is an Export-Processing Zone (EPZ) type of industry, which has helped to boost the economic development of Mexico from the 1960s in an effort to implement import-substitution industrialization by industrializing the Mexican-US border. The *maquiladora* industry is at the heart of globalization by being on the border with the United States, squeezed between Latin American countries in the process of economic development and North American countries in quest for further growth opportunities. As a consequence, the case study of its workers presents a vivid opportunity to study the interdependence of opportunity sets between multiple identities, geographically, culturally, and historically, which will be explored further in the next chapter. For now, the second section of this chapter presents the external shocks on a *maquiladora* worker's E-mapping from the time of implementation of the North American Free Trade Agreement in January 1994. Finally, the last section exposes the theoretical consequences of external shocks on the economic entitlements of the *maquiladora* worker's identity, namely, real wages, which represents the purchasing power of workers, and the demand for labor in that industry, which represents the employment opportunities for current and potential workers.

4.1. External Shocks on Identity E-Mapping

Entitlements, or access to goods, services, and potential functionings, are by default dependent upon others' entitlements. For example, access to a given commodity by one individual, assuming that it cannot be shared in consumption, restrains access to this commodity by another individual. The entitlement set of an individual starts at the household level, since the household represents the first environment where an individual learns to function in relation to others. From childhood, an individual is thus entitled to, or has access to, commodities and potential functionings according to personal conversion factors of commodities into functionings, together with social and environmental conversion factors, including identity-related norms of behavior. In this sense, Comim (2008) argues that most capabilities can be described as "social capabilities." Capabilities therefore depend on the opportunities faced by individuals in a living environment, and also by the way these opportunities can be affected by the social and environmental conversion factors of commodities into functionings. From this perspective, a potential functioning is the result of the interaction between access to commodities and the way the multiple identities of the individual concerned are perceived by oneself and by others within the household and in the society as a whole. E-mapping in Sen (1981) is defined as the function that specifies the set of alternative commodity bundles that the person can command, respectively, for each endowment bundle. In other words, E-mapping represents the set of consumption bundles that the individual faces, any of which can be chosen, given his or her endowments. However, the methodology used here is to include both commodities and functionings in identity E-mapping in order to look at economic events at the level of a group identity instead of a traditional perspective of looking at more aggregated levels of analysis. In effect, studies on the impact of economic events and policies tend to use cross-country analyses,[2] and they also tend to focus on the economic impact at the national level[3] or the industrial level.[4] Similarly, studies related to household welfare concentrate on changes in income or in the income distribution without naming other domains that could have been affected by economic events or policies.[5]

External shocks on an identity's choice set could either reinforce this identity over time or create entitlement failures for those weakened by the shocks. Entitlement failures are characterized as individuals' reduced freedom to choose in terms of restricted alternatives due to one or more of their identities. From the individual's point of view, the impact of economic events on identity well-being is best understood by looking at the living environment he or she is entitled to, that is, the nature of social and economic entitlements specific to each individual's set of identities. To understand the rationale behind this, let us investigate the way in which market opportunities and external shocks may affect capabilities through identity E-mapping. Let us assume an individual with equal endowments in two different countries, A and B, which display two different sets of social and economic entitlements from time t_0 to time t_1. From time t_0 to t_1, country A experiences a high growth rate of GDP, which potentially expands the economic entitlements of the individual in that country. The potential increase in personal endowments at time t_1 will depend on the functioning of personal identities relative to others. For example, if the norm of fairness in country A is that a dominant identity should be entitled to all resources, then other identities are regarded to be dysfunctional elements for that established norm. On the contrary, over the same period from time t_0 to t_1, country B is hit by a tsunami, which damages the economic entitlements of all individuals in that country, thus lowering individual endowments at time t_1. If the norm of fairness in country B is that a dominant identity should be entitled to all resources, and assuming that there are no resources remaining, then the only certainty remaining is to function as a human being. As a result, the human identity should trigger cooperation between individuals to make sure that functionings as a human being work, regardless of other personal identities.

To go deeper into that issue, one needs to look at the mechanisms involved within each identity and across identities. Two phenomena identified by the literature in both economics and psychology can be linked to identity-utility ideals to investigate the effects of external shocks on identity E-mapping. First, an external shock on identity E-mapping should trigger adaptive preferences or habit formation toward identity-related ideals. Second, an external shock on identity E-mapping should trigger interdependent preferences or a social comparison.[6] First, adaptive preferences underline the fact that individual material aspirations change over time as people tend to get used to their endowments, thus putting a higher value on their endowments at time t than at time $t - 1$. From the point of view of the literature in psychology, this notion could also be related to the fact that individuals experience the "endowment effect" (Huck et al. 2005) or "treadmill effect" (Kahneman 2000). Both effects represent the notion that people have a tendency to get used to their endowments over time. In that sense, people adapt personal perceptions and judgments of the items they own according to a socially constructed benchmark or optimality point. Therefore, one could argue that adaptive preferences exist because people experience endowment or treadmill effects. Here, two aspects of preferences need to be clearly separated. The first is the general notion that individual preferences are molded by the society in which the individual lives. This is what Bowles (1998) calls "endogenous preferences," and this is part of an individual's social entitlements in this manuscript. The second aspect is the individual evaluation of his or her endowments, which changes with the experience of that level of endowments. This refers to adaptive preferences.

Markets and other economic institutions influence tastes, values, and personalities; the concept of endogenous preferences is therefore a natural feature of human behavior. Endogenous preferences mean that individual preferences are partly shaped according to the organizational structure of the market (Bowles 1998). By economic institutions, Bowles refers to rules of market behavior that differ in their organizational structure, such as corporatist, capitalist, traditional, communist, or patriarchal societies. A communist organizational structure may shape preferences in a way that is different from a capitalist organization. A corporatist, communist, or traditional society is based on social values that then lead to a system of economic organization. Under such circumstances, the boundary between economic institutions and other institutions is not clear; values within an organization, an industry, or an institution depend on the values of the wider society. However, Bowles's argument highlights the fact that markets influence cultural values. In many countries, for instance, the intensive marketing and supply of beverages from the Coca-Cola Company has affected consumers' tastes up to the point where water is becoming a secondary drink on kitchen tables. Markets and economic institutions are part of an individual's economic and social entitlements. To a certain extent, it seems that endogenous preferences relate to the concept of adaptive preferences to material aspirations or hedonic adaptation. Institutions create a sense of cultural identity that sets the benchmark of how social relations should ideally be, and people then learn to function accordingly, that is, individuals adapt to an institutional ideal.

In terms of living standards, adaptive preferences come into action once a certain level of income is reached, which looks for all countries to be around $15,000 per capita in 2000 US prices (Layard 2003). This benchmark comes from a measure of subjective well-being by level of economic development. Satisfaction with life as a whole from the World Value Survey and GNP per capita from the World Bank were plotted together for 65 countries in 1995 (Inglehart and Klingemann 2000). The results described by Layard must be interpreted cautiously, given that they are based on a single year. In effect, historical and country-specific characteristics need to be taken into account. For example, ex-Soviet countries are at the bottom left hand corner of the graph, suggesting that they were among the countries with both the lowest income per capita and with people who were least satisfied with life as a whole. From its historical context, this fact might be related to the fall of the USSR and the Berlin wall in 1989, leading to the sharpest drop in real incomes over the five years preceding the data collection. The political climate may be one of the main reasons for low life-satisfaction among countries such as Ukraine, Belarus, Russia, and Lithuania. Apart from the ex-Soviet countries, however, the correlation between level of satisfaction and level of income appears positive up to an income level of $15,000 per capita. This suggests that once basic living needs are satisfied—depending on prevailing income levels—adaptive preferences will make the accumulation of commodities insignificant with regard to life satisfaction level. Barr and Clark (2007) have illustrated the adaptation hypothesis—whereby individual aspirations adjust to current living conditions—in three communities of South Africa looking at income, educational, and health aspirations. Their findings are consistent with individuals' adaptation to current living conditions in relation to a reference group and past living conditions. Here, the hypothesis of group identity–related ideals setting the benchmark of living standards seems plausible.

Another view of adaptive preferences is in the literature on happiness, where Easterlin (2005) argues that people tend to go back to their "set point of happiness" unless some powerful events make them depart indefinitely from this point. Different preferences are applied in different situations in a manner that need not be consistent through time as long as the individual returns to this set point. This, in turn, depends on people's hedonic quest. Hedonic quest means literally that the quest refers to pleasure or the minimization of unpleasant feelings, whether it involves altruistic or individualistic behavior. At a point in time, if people have preferences for altruistic behavior or individualistic behavior, the choice between the two alternatives need not to be rational, that is, repeated consistently through time. Individuals can act in an individualistic manner in some circumstances and altruistically in others as long as they can sustain their happiness and shorten unhappy feelings.[7] However, the argument put forward in line with the E-mapping framework is that hedonic quest and the "rational" behavior attached to that quest tend to be linked to identity-related ideals and how far identity-utility is from that ideal. A mother might be altruistic toward her children, which would be part of her identity-utility as a mother who has learned to function with her children. A manager's quest for profit maximization might be related to a status-seeking behavior in consumption, which would be part of her identity-utility as manager but which could equally be linked to a breadwinner type of behavior in her identity-utility as household member. Similarly, a volunteer in a nongovernmental organization (NGO) may seek altruistic ends in order to be closer to identity-related ideals as a member of an NGO defending a cause he has reason to value.

Utility maximization becomes important in well-being measurements once it is understood toward which optimality point this program is directing. Considering utility maximization out of income, as status-seeking consumers, for example, misses out the complete picture of an individual as a combination of identities and the potential effects of economic events on each identity. Whether the effects of an economic event are positive or negative, individuals will seek the least negative outcome for their own well-being by trying to go back to their set point. For instance, assuming that currency devaluation hits a country, inflationary pressure on purchasing power will lead people to seek the highest and/or most secure income possible after the devaluation. Individuals and households seek higher money wages to offset higher prices in order to maintain their living standards. Based on the British Household Panel Survey and using time series of income satisfaction over a ten-year period from 1990 to 2000, Burchardt found that income is a poor proxy for measuring utility, utility being measured by self-reported levels of income satisfaction, thus ignoring other factors that may influence satisfaction. People who have experienced a previous fall in their income are less satisfied than people whose income has been constant in real terms (Burchardt 2005). However, people experiencing increasing income over time are not more satisfied than those having a constant level of income. People suffer more from falls in income than they gain from rises in income.[8] Burchardt (2005) has regressed income satisfaction, an ordinal variable, on current log income and various control variables using cross-sectional ordered logit. The regressions are made in terms of changes in income and according to different levels of income. Therefore, one important conclusion to highlight is the role played by the previous level of income on satisfaction levels, but only in terms of a fall in income. Once an optimal living standard is reached, it is difficult to lower this standard, whereas increasing this standard will not affect income satisfaction much.

In the event of external shocks negatively affecting the individual's access to income, living standards will be assessed by the individual comparatively with other individuals sharing or not sharing an identity, which is where interdependent preferences are triggered. Interdependent preferences relate to the interaction between individuals in the sense that others' living standards influence one's preferences for one's own living standard. The endowment effect and adaptive preferences predict that the accumulation of commodities leads to higher living standards but has no significant effect on happiness levels since people tend to go back to their set point. However, regarding market interactions, interdependent preferences can have a perverse effect on people's level of satisfaction. Based on Danish time-use survey data, Bonke (2005) showed that the higher the man's share of the total household income,[9] the higher his economic satisfaction. This is largely in terms of the man's labor market earnings relative to the woman's earnings, rather than in terms of the share in expenditure decisions. In other words, it relates to the perception of a male identity, that is, what a man should do or should be, vis-à-vis the perception of female identity, that is, what a woman should do or should be. By economic satisfaction, Bonke means subjective satisfaction with individual economic status both in terms of employment and relative income. Similarly, "men's satisfaction with their economic status declined significantly when they had the opposite employment status (employed or unemployed) to that of their wives" (Bonke 2005). Whether the effect was similar for women is not reported, but it certainly raises the issue of relative income and interdependent preferences as a determinant of human satisfaction. It means that given the income received by men within the household, they will react according to the income received by their female counterparts. Here, the undertaking of paid work by women is not seen as an unsatisfactory outcome for men as long as the female monetary contribution to the household is not greater than the male contribution. The point here is not simply about the terms of labor supply decisions, but rather that market opportunities for male and female potential and current workers play a major role in sustaining the existing cultural norm.

External shocks affect an identity's economic entitlements through both adaptive and interdependent preferences within and across identity groups. Economic entitlements deal with the employment opportunities, given individual's skills; with the access to commodities, goods, and services, given a monetary income and a certain level of prices; and with the access to goods and services provided by the state, when applicable, given the individual level of endowments. Including both commodities and functionings in the identity E-mapping is a means to accurately account for individual skills and other personal characteristics in well-being measurement and access to potential functionings. Taking the example of command over corn, Sen (1981) proposes an illustrative model of exchange entitlement with two people. The exchange entitlement of individual is illustrated as a ratio of the personal monetary income over the price of corn. However, it seems reasonable to think of the last example as a limited case of exchange entitlement where the individual can exchange a part of what he owns (money income) for a certain amount of corn, given its price p. Other factors, such as the level of prices members of the household face as consumers, job opportunities they face as workers, or their level of well-being according to each individual's set of identities, are also affected by economic events and policies. According to an individual's set of identities (consumer, employee, household member, to cite a few examples), the channels through which those policies act on capabilities become different from another individual's set of identities. For instance, in the event of currency devaluation, the owner of an export-oriented firm is better-off in terms of profit maximization due to the rise in the demand for exports. However, a wage earner from the same firm is worse off in real terms, provided that nominal wages do not follow the rise in the price level. As a consumer, the same wage earner will face higher prices due to devaluation. As a household member, the same individual has to manage the lower purchasing power to, preferably, achieve the same standard of living. His or her preference for maintaining the same living standard comes from an aversion to cuts in purchasing power according to a certain benchmark. Studying policy effects through E-mapping thus means that the analysis starts from a specific identity, which in this case is the *maquiladora* worker's identity.

4.2. External Shocks on a *Maquiladora* Worker's E-mapping

The economic integration post-NAFTA led to an increase in Mexican real GDP by 25 percent over the period from 1995 to 2000 (INEGI 2011). However, under the heading of NAFTA's effect on Mexican economic growth hides a much more important factor that has contributed to this economic growth: the peso devaluation of December 1994. One argument has indeed been that the impact of NAFTA on the *maquiladora* industry has not proved to be significant, given that the industry was already established for more than three decades with duty-free import and exports (Gruben 2001). Several other economic events hit the *maquiladora* industry since 1994, including the US recession in 2000–2001 and increasing competition from Asian exports of labor-intensive products. In November 2001, for example, China joined the WTO, thus getting access to wider market opportunities and putting pressure on its competitors, notably those in Export Processing Zones. China's entry into the WTO put pressure on *maquiladora* workers' wages, since China was seen as a threat to *maquiladora* output, mainly because of the scale of this entry's affect on world competition compared with the competition from other Asian countries such as Vietnam or Cambodia.

The 1994 peso devaluation contributed to a rise in the demand for Mexican exports, and in the demand for *maquiladora* output in particular, which accelerated growth in the industry. Figure 4.1 shows the sudden rise in Mexican Consumer Price Index (CPI) following the peso devaluation. From 1995 onward, the CPI rose at a speed that was never experienced before over the period 1969–2009. This sudden and fast rise in consumer prices may have had tremendous repercussions on the rate of human development of its population. Potential consequences could have been either an acceleration or a deceleration of exchange entitlements that affect negatively or positively the individual and collective functionings of its population. The next chapter investigates the consequences of this external shock on the collective functionings of the Mexican identity over that period, that is, changes in social entitlements of the *maquiladora* identity in terms of health, education, migration flows, and so on. For now, the focus will be on the effects of external shocks on the main economic entitlements of the *maquiladora* identity, namely, real wages, which represent the purchasing power of workers, and the demand for labor in that industry,

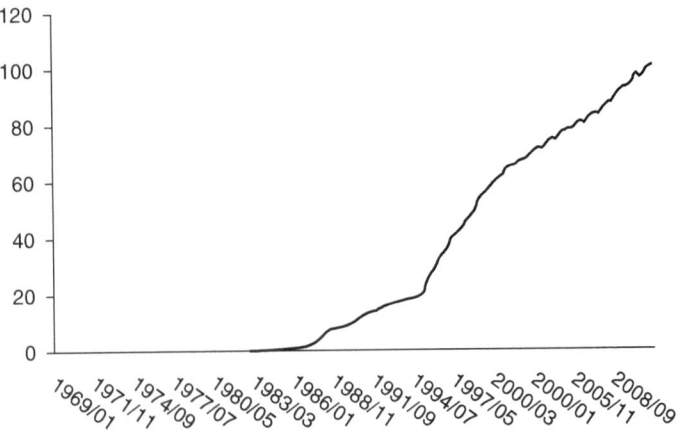

Figure 4.1   Consumer Price Index (1969–2009)
Source: Banco de México (via INEGI 2011).

which represents employment opportunities for current and potential workers.

To comprehend the effects of economic events on the *maquiladora* identity, one needs to understand the channels through which those effects might occur, namely, the consumer channel and the producer channel. First, through the consumer channel, currency devaluation leads to a rise in the level of prices, which puts pressure on consumers' income and purchasing power. In other words, inflation decreases consumers' real wages so long as nominal wages do not follow the increasing trend of the price level. In the example of Mexico, the peso devaluation of December 1994 was followed by an increase in the general level of prices by 50 percent for the year 1995.[10] This level rose constantly, reaching a 300 percent cumulative increase in the price level from 1995 to 2002. Inflationary pressure on household income is one of the channels through which poverty can easily deepen, especially when wages do not keep up with the increasing level of prices. In terms of subjective well-being, Frey and Stutzer (2002) surveyed the empirical literature on the effect of inflation on people's level of happiness. Their conclusion is that this effect is negative whether or not inflation is predicted. "People resent inflation, not least because they fear a lowering in their (future) standard of living, a worsening in the distribution of income (in particular, that some people are exploited), the moral effects on the cement of society, and the political and economic unrest produced, as well as concerns about a loss of national prestige" (Frey and Stutzer 2002, 115).[11] Consequently, it seems reasonable to support the view that the sudden rise in the high level of inflation makes people's level of happiness deviate from their "set point" in different identity-related ideals without necessarily returning to this point in the long run. For example, the loss of national prestige is related to cultural identity. Equally, the loss of purchasing power is related to consumer identity. And finally, lower real wage may also affect their wage-earner identity, which could potentially lower the level of effort and productivity. Second, through the producer channel, the effect of currency devaluation makes the country's exports cheaper through exchange rate mechanisms. Therefore, a higher demand for the country's goods and services triggers the local firms' production and labor demand in export-led industries. In both cases, the exchange entitlement of an individual working in that particular industry will be changed, first, through the access to a nominal income that has to be spent given a particular inflation rate, and second, through the demand for a specific type of job that can be affected by the change of the industry's structure as a consequence of economic events. Then, these two channels will have repercussions on the role and position of the working individual within the household. As was discussed in chapter 2, social norms have an impact on both channels. Cultural and gender norms influence the household decision concerning the labor supply of individual members, while norms of fairness influence cooperation in market interactions, price and wage setting, and labor demand decisions.

Additionally, since social norms are based on perception, that is, observed and, therefore, limited regularities, the speed of adjustment of perception to economic changes often lags behind the speed of market dynamics. When thinking of money as a social convention,[12] exchange rate fluctuations show the extent to which social preferences can negatively affect the value of money: a drop in confidence in a currency as a store of value can easily spread wide through adaptive and interdependent preferences, leading to a drop in the market value of the currency. Similarly, perception of monetary value leads to money illusion, which plays an important role in shaping individual expectations of price changes. Individuals have difficulties in readily translating the observed nominal values into real values. For example, in the short run, workers tend to respond more quickly to nominal wage changes than to price level changes in deciding whether or not to quit, which Carter (2005) attributed to workers' perception of fairness. The speed of reaction to wages and prices may be linked to people's perception of fairness relative to others' wages, or that the change in nominal wages is perceived as a change in prices. Shafir et al. (1997) surveyed different theoretical and empirical works on money illusion, mainly in terms of the discrepancy between real and nominal wages. People's perception of the general price level lags behind actual events, such that when inflation begins to rise, it takes people some time to realize it fully, and they therefore tend to underestimate the price level. This means that the perception of monetary values is readjusted in real terms after a certain time lag.

Economic growth in a market economy concerns the allocation of scarce productive resources to produce an ever increasing range of goods and services. Once productive resources are exhausted, additional productive resources from the household are pulled into the market in order to maintain the same standard of living, especially in a context of rising inflation. In order to sustain the same purchasing power, previously unemployed labor from the household will have to be pulled into the labor market. This argument is in line with López et al. (2008), who commented on the features of the Mexican labor market, such as the lack of unemployment benefits and low levels of family income, "which does not enable the family to support its unemployed members" (19). Both the institutional factors contribute to the sustainability of a low rate of unemployment and a high rate of informal employment, around 4 percent for the former and 28 percent for the latter in the second half of the 2000s (INEGI 2011). In the context of the individual E-mapping of a *maquiladora* worker, macroeconomic events have affected the economic entitlements of male and female *maquiladora* workers differently due to their different social entitlements. Real wage and employment are amongst the major economic entitlements[13] of *maquiladora* workers. Real wage reflects a worker's purchasing power. A job position, having an associated wage with it, and the general price level are the determinants of purchasing power. Consequently, any economic event leading to the impoverishment or empowerment of people's living standards has to go through these two elements: employment and real wage. The next section highlights, theoretically, the factors determining the real wage and employment of *maquiladora* workers to assess their degree of interaction in the event of external shocks on these two economic entitlements.

4.3. Consequences on Economic Entitlements

By identifying the link between short-run fluctuations and long-run movements in the real-wages–employment relationship, the factors affecting the worker's economic entitlements and subsequently the worker's capabilities can be understood and explored further, which is carried out in the next chapters. For now, this section gives an overview of the theories behind the wage-employment relationship. Operating under imperfect competition, an industry sets its wages after considering the demand for the industry's output, which is affected by macroeconomic events.[14] The extent to which these events can be detected empirically depends upon whether a short-term or a long-term perspective is adopted, that is, the resulting movements in the real-wages–employment relationship may have different determining factors according to the perspective adopted. This overview therefore starts by analyzing the factors influencing the relationship in the short run and in the long run before presenting the problems faced when examining it from a dynamic perspective, namely, the problem of simultaneity between the supply of labor and the demand for labor, and the problem of norms and expectations in market interactions.

Sign of the Wage-Employment Relationship

Under the theory of perfect competition, the relationship between employment and wages is clear. It has to be a negative relationship, from one of the first-order conditions for profit maximization. Indeed, under the assumption that the firm operates where the marginal product of labor is declining (reflecting the first-order condition), the labor demand curve can be derived from the production function and the demand for the firm's output as a downward sloping curve showing a negative relationship between real wages and the level of employment.[15] However, the general consensus is now to adopt imperfect competition as a more accurate representation of the actual market situation. Under imperfect competition, the relationship between employment and wages can be positive or negative. It is considered to be negative when the firms are operating with a diminishing marginal product of labor (which is assumed to always be the case under perfect competition). Looking at the industry level, however, requires a broader understanding of the factors affecting the real-wages–employment relationship.

At any time, economic events can hit an industry. These events are then translated into short-run adjustments of the capital-to-labor ratio and other adjustments, such as changes in the level of output, output prices, and nominal wages. Over time, this can lead to a change of industry structure, from a labor-intensive output toward a more capital intensive output, for example. The resulting effect on an individual decision-maker at the firm level is that the firm may have difficulties facing these changes and not having many options to choose from. The firm may either choose to exit or to seek to lower its costs, but the strategic decision will be reflected by the resulting effect on the capital-labor trade-off, given the wage level set in part at the industry level and according to local unemployment. The change of structure, taken as an example above, is a consequence of firm-based entry or exit decisions. However, the change in the balance of capital to labor that affects the structure of the industry is better detected with industry-based data, because firm-based data show mainly the entry and exit or investments related to the industry's change of structure. The last chapter of the book gives a taste of the structural change of the *maquiladora* industry per sector and shows the extent to which changes in labor-demand affected the gender composition of the industry by sector.

From cross-country and cross-industry evidence, the employment-real wages relationship has been seen as a positive factor, especially in the context of the wage curve (Blanchflower and Oswald 1994). The wage curve as defined by Blanchflower and Oswald (1994) refers to the cross-section evidence that there is a negative relationship between wages and the local rate of unemployment. The elasticity is found to be around minus 0.1. Nijkamp and Poot (2005) revisited the elasticity and found it to be approximately minus 0.07, with important variations across the studies surveyed. In other words, the higher the local unemployment, or informal sector employment in the context of developing countries, the lower the bargaining power of workers to negotiate for higher wages. A negative relationship between unemployment and wages can also be interpreted as a positive relationship between employment and wages, which means that efficiency wage theories become highly relevant to explain this relationship. Indeed, the positive relationship between employment and wages is consistent with bargaining and efficiency wage models.

The modern approach to efficiency wages means that employers set wages above the market-clearing wage. Employers may wish to do so in order to reduce costs of labor turnover, to reduce shirking by employees (Shapiro and Stiglitz 1984), and to increase labor productivity while saving on total labor costs (Carlin and Soskice 2005). Also, one may argue that the benefits offered by the *maquiladora* industry are part of the shirking approach, such as providing means of transportation to the workplace to prevent absenteeism. However, the "shirking" point of view provides no clear explanation for discriminatory practices and wage gaps between groups of workers in the same job position. In contrast, the fair-wage-effort hypothesis (Akerlof and Yellen 1990), with social norms attached to the concept of fairness and identity, may explain this wage differential. Adopting this approach will also enable us to include both the "gift-exchange" approach (Akerlof 1984; Akerlof and Yellen 1986) and the "shirking" approach as particular cases of a broader fair-wage approach with evolving social norms. For example, the norm of fairness may differ depending on the perspective adopted: whether fairness is understood from the employer's point of view or from the employee's side. In the shirking approach, employees may deem it fair to shirk on the basis that the value of their effort is not reflected by their wages. In other words, identity-utility as an employee is not maximized according to a perceived ideal, which, for example, could be that wages between employers and employees should be equal. Similarly, in the gift-exchange approach, the wage-effort relationship is perceived by both employers and employees as positively correlated: it is fair to provide more effort if the reward for providing this extra effort is higher.

Causality of the Wage-Employment Relationship

Under the neoclassical approach, in which wages are set according to market mechanisms, at the level of the firm the real product wage determines the amount of labor employed. Therefore, depending on the real wage, the firm will hire a certain amount of labor. If wages increase, then jobs need to be cut, and vice versa. Under the Keynesian approach, aggregate demand determines employment, and the resulting marginal cost and marginal product of labor set the real wages. Assuming imperfect competition, the real product wage and employment are set together by the decision of the firm in the context of the level of demand that it faces. Enterprises decide to operate for one particular level of demand; a different level of demand would lead to a different choice of real product wage and employment (Sawyer 2002). Assuming that the wage is set at the level of the industry, with regional variations according to the local rate of unemployment, the firm has to choose the optimal level of employment under the constraint of wages and according to its amount of capital, and to the level of output demanded. However, problems of externalities, asymmetric information, menu costs or expectations hinder the accurate perception of this level of demand. Thus, several problems lead to a complex picture of the employment-wages relationship.

In the context of the *maquiladora* industry, the causality between wages and employment is investigated in relative terms in the context of the gender identity of *maquiladora* line workers. This means that the economic entitlements of the *maquiladora* identity presented in the last chapter will be looking at the significance of labor costs, both female and male labor, in determining changes in the level of male and female employment over time. In the short run, it is expected that firm-based factors such as capital, inputs, or wages will play a significant role in explaining the level of employment. Short-run adjustments reflect the changes in the capital-to-labor relationship in terms of the real product wage resulting, in the long run, from changes in the demand for the industry's output. Looking at the industry level, external factors such as exports or the exchange rate are argued to be more easily detected in the long run. Therefore, it is assumed that real product wages are determined at the industry level and that they influence the level of employment. In the long run, however, real product wages are secondary in explaining the level of employment, since external macroeconomic events determine the level and structure of the industry output demanded. In other words, in the long run, the industry price is driven by the influence of external events on the demand for the industry's output, which in turn affects the real product wage.

Problem of Simultaneity in Supply and Demand

Another problem exists in examining the employment opportunities of potential and current workers, and that is the simultaneity between demand and supply sides of labor, especially with a time perspective. In the short run, the profit-maximizing firm will hire workers until the point at which the marginal cost of labor, that is, the cost of hiring in real terms, equals the marginal revenue of labor, that is, the change in output resulting from hiring an additional worker, holding the quantities of other inputs constant and given the level of demand for the firm's output. The two conditions, that other inputs are held constant and that the price of output is given by the market forces of supply and demand, tell us that it is a short-run perspective. Assuming that there are two inputs, labor and capital, labor will generally have a marginal product greater than zero if capital is held constant (Ehrenberg and Smith 2006), but it will still decrease as employment rises. In the long run, when firms are able to vary their capital stock, the marginal product of labor is still the key element to determine the labor demand curve. However, other conditions change: other inputs are not fixed and the firm has to choose the amount of capital and labor to achieve a certain level of output. In general, the demand for labor is more elastic in the long run than in the short run, since both substitution and scale effects are in action.

Translated to the industry level for the *maquiladora* identity, wages are set in an imperfectly competitive environment at the industry level, given the demand for the industry's output.

In the short run, and according to neoclassical theory, wages are the primary determinant of the industry level of employment. The main assumption under imperfect competition is that prices, and thereby the real product wage, are set to maximize profits, given the conditions under which the firm operates, including the level of demand, the events affecting this level, and the norm of fairness behind market decisions. In the long run, wages are exogenous to the employment level in the sense that macroeconomic events influence the demand for the industry's output, which includes the structure of the industry output or the capital to labor ratio: the resulting effect on wages identified in the short run is therefore hidden away by the long-run perspective. Adopting a dynamic model that looks at the changes in employment over time is innovative in this area of labor economics, since most empirical work relies on cross-industry or cross-firm data and rarely focuses on changes in employment and wages over time within a single industry, and within a single job position of line worker with two different identities, male and female. Hamermesh (1993), for example, conducted a thorough survey of all the different empirical works on labor demand. According to his survey, several alternatives have been used, such as using supply and demand equations in a simultaneous equations system.

Out of 70 studies conducted, Hamermesh showed that most studies produce estimates of the own-wage elasticity of labor below one, which is based on the assumption that causation runs from wages to employment. In other words, the demand for labor seems inelastic to a change in wages. However, once real wages are included and once the causality is reversed and runs from employment to real wages, the relationship may be negative, zero, or positive depending on the demand for output (Dunlop 1938; Michie 1987). In chapter 3 of Michie (1987), a survey of the empirical literature on earnings and business cycles does not lead to a clear vision of the sign of this relationship. One reason for this is related to the definition of a business cycle, which varies across studies and across the different experiences of countries. At the industry level, the argument follows from Robinson's description of the industry's curve for labor as the curve of average net productivity and that "the analogy between the competitive demand curve for labor and the competitive supply curve of the commodity is very close" (Robinson 1969, 253). This means that the industry will supply output according to the demand for its output, which in turn will determine the level of wages.

In the case of the *maquiladora* industry, the changes in labor supply will be investigated in the next chapter on the social entitlements of the *maquiladora* identity, in which changes in social entitlements are assumed to be dictated by the economic entitlements of workers. For example, in the context of an economic downturn (boom), a larger share of the household labor supply is expected to enter the labor market in order to sustain (increase) the household's standard of living, or as a consequence of economic pressures (opportunities) that could create tensions and split households. For example, divorces in Mexico increased by 65 percent over the period from 1990 to 2006, while female-headed households increased by 75 percent (INEGI 2011). Meanwhile, the fertility rate dropped from 3.4 children in 1990 to 2.2 children in 2006, which resulted from an increase of the female labor participation, from 36.9 percent in 1990 to 41.5 percent of the working-age population in 2007. One factor suspected to have shifted the effective labor supply is the same factor that is suspected to have shifted the labor demand in the *maquiladora* industry, namely, the 1994 peso devaluation. Therefore, the issue is to identify the magnitude of the shifts in demand for and supply of labor. On the supply side, inflationary pressures on households due to the devaluation led more household members to ask for a market job, such as a *maquiladora* job. On the demand side, the depreciation of the real exchange rate caused by the currency devaluation led to a sudden rise in the demand for the industry output and therefore a rise in the demand for labor to produce the additional output. The attractiveness of Mexican exports following the devaluation together with NAFTA's implementation led to a real GDP growth of 25.3 percent over the period 1995–2000, followed by a slowdown, with a GDP growth rate of 6.5 percent over 2000–2005, possibly linked to the US slowdown (INEGI 2011). In the light of these economic changes, the labor supply is assumed to constitute a reserve army of "discouraged work-ers" who cannot be hired at the prevailing real wage (Sawyer and Spencer 2010). Indeed, the working population in that industry is mixed and can be drawn from different parts of the labor force: from the unemployed; the informal sector; the young unskilled population with a high turnover; or from more mature female and male workers, as described above, in light of the increase in divorces over the past 15 years. Thus, employ-ment is in effect determined by demand and is usually less than the available supply.

Problem of Norms and Expectations

Norms represent the way people perceive themselves and oth-ers, as to how they *should* behave and how others *should* behave. In a lecture given at the American Economic Association in January 2007, Akerlof highlighted the fact that norms are economically consequential. Due to the role of norms in con-sumption behavior, investment behavior, and wage and price behavior, he succeeded in breaking up the five neutralities accepted in macroeconomic theory—namely, the indepen-dence of consumption and current income, the independence of investment and finance decisions, inflation stability only at the natural rate of unemployment, the ineffectiveness of stabi-lization policies with rational expectations, and the Ricardian equivalence. Indeed, the search for positivism in economics has led to many frustrations due to the subjectivity of human behavior, which is difficult to model.

Expectations or perceptions of the future depend not only on the economic environment but also on the social and cul-tural entitlements from which those perceptions have been formed. Expectations and norms have important implications on labor demand through, for example, short-run sticky wages or perceptions of fairness, as developed in the second chapter of this book. For example, a climate of fear and dictatorial lead-ership certainly will not help to see the future as a bright one and to stimulate entrepreneurial spirits (Djankov et al. 2003). Therefore, expectations nest norms and ideals of consump-tion behavior, investment behavior, wages, and price behavior.

A common critique of mainstream theories has been that firms' managers do not have sufficient knowledge and patience to enter into complicated mathematical equations to calculate the marginal product of their labor or to be a profit-maximizing firm by equating x with y. A response to this argument has been that managers will behave "as if" they were profit-maximizing/cost-minimizing firms, since they would otherwise not survive in a perfectly competitive world. As identified in the context of the E-mapping framework, managers may have objectives other than profit maximization due to the multiple identities to which they belong. The search for personal development, social status, and constant or increasing household living standards are a few of the reasons behind entrepreneurs' behavior for business success or sustainability. In reality, managers make strategic mistakes or take decisions based on asymmetric information, which bias their expectations of the output level. They have expectations about the expected demand for their output, the economic climate and the risk for investment associated with it, and the future level of employment needed in the light of the two previous factors.

From the perspective of the real-wages–employment relationship, employers have a set of norms, based on their unique set of identities, concerning the individual they wish to employ, as well as the level of relative wages they wish to set depending on the group of individuals they hire. In turn, those norms are sustained by adaptive and interdependent preferences. Adaptive preferences at the firm level mean that in the event of wrong expectations about price levels or employment levels, a period of adjustment is required for the decision-maker to adapt to the reality of the demand for the firm's output. Those adjustments will then be made according to the norm of fairness of decision-makers, which are strongly influenced by identity-related ideals of how prices and employment should be decreased or increased. Interdependent preferences mean that the firm's strategic decisions are made in the light of the decisions made by competitors. Therefore, it is likely that norms applied in one firm are consistent across the industry if competitors share similar identity-related ideals. For instance, in the context of the demand for labor, provided that one type of worker is seen as more suitable for one particular type of job than another type of worker, after controlling for individual skills and abilities, it means that there will be discriminatory practices in the hiring process against the second type of workers from one firm to another across the industry.

The problem of perception in norms and expectations influence the decision-making process and the strategic behavior of entrepreneurs. Entrepreneurs have expectations about the future level of demand, as they have certain norms about the price level they should set, which lead to short-run fluctuations. When taking into account the problem of norms regarding individual behavior—whether it concerns the household or the firm—it seems sensible to talk about bounded rationality (Conlisk 1996) anchored in the problem of imperfect information. For instance, decision-makers need to find a way to assess the marginal cost and marginal product of labor of the workers the firm wishes to hire. The information available to establish the cost of that new worker is the industry average salary given for a specific job. This average salary is argued to be in part based on the social perception of the type of worker being hired, that is, depending on the social value of the effort-related capabilities attached to that type of worker. If an immigrant worker, a nonimmigrant man, and a nonimmigrant woman apply for the same job position, the social perception of the immigrant worker, a man or a woman outside the firm, is likely to influence both the hiring decision and the level of wages set and serves as a basis for discrimination.

4.4. Conclusion

The wages-employment relationship presents a complex picture depending on the level of analysis adopted: depending on whether it is from a short-run or long-run perspective, from the perspective of employers or employees, or from the perspective of the firm or the industry. When the labor supply side thinks in terms of real wages, that is, purchasing power, the labor demand side thinks in terms of real product wage, that is, the actual quantity of output produced. The firm has to adopt a real-product wage/employment combination according to the level of demand it faces. The industry sets efficiency wages, which leaves the firm able to adjust the employment factor and relative labor costs given expectations of aggregate demand. If employers behave according to the cost-minimization program, they will choose the type of labor force and set their labor cost according to the information concerning the type of labor force they wish to hire. The available information strongly relies on their perceptions of the workers outside the firm, that is, the social perception of the worker's set of identities: whether male or female, young or old, indigenous or of Spanish descent. Consequently, the impact of external shocks on the *maquiladora* worker's E-mapping will depend on other identities of the *maquiladora* line worker. Considering an individual to be a unique set of identities, the analysis needs to go further into the interdependence of identity E-mappings for a *maquiladora* line worker. A *maquiladora* line worker can be a Mexican male or female, which requires looking at different E-mappings, namely, the E-mapping of a *maquiladora* worker, of a Mexican man, and of a Mexican woman, within the labor market and within the Mexican society.

The entitlement approach to the *maquiladora* worker's identity developed in chapters 5 and 6 considers the economic and social opportunities faced by one group of individuals sharing the common identity of *maquiladora* line worker in Mexican society, but with a different gender identity. Starting from each identity allows us to understand the relative changes in the socioeconomic environment of these workers, instead of establishing an absolute level of entitlements for a particular identity. Thus, the economic and social entitlements include access to a job position with all the hardships and responsibilities that it poses (line-worker, long working hours in a standing position, productivity targets, and so on), income and work benefits, access to social services, social norms, and the legal framework. Depending on their role in the household, it appears that the gender gap in the industry is sustained and fed through the role played by women within the household as mothers, daughters, sisters, sisters in law, or any other identities. With the instance of the gender gap in the *maquiladora* industry, it can be shown that social entitlements are built in such a way that the difference in economic entitlements affecting the household unit is translated into a wage gap effect in the labor market. From the point of view of the firm, there is no discriminatory practice, in the sense that a lower income entitlement for women is already socially accepted. The general social acceptance means that the "going wage" for women is less than that for men, and the firm accepts this "going wage." In such circumstances, increasing market competition does not reduce those discriminatory practices when changes in social entitlements could. Simultaneously, social values within the household are affected by the pulling of resources into the market economy; fertility rates decline and more financial resources are put into education of children and their care outside the household. The point here is to show the extent to which economic events affect an individual's opportunities to achieve potential functionings and to function in a living environment according to identity-related ideals.

Chapter 5

Social Entitlements

In 1994, Mexico experienced two major economic events: the NAFTA implementation in January led to the free movement of goods and services, but not of labor, across the US-Mexico border, and the peso devaluation in December led to a severe recession in 1995. The extent to which these changes affected the well-being of *maquiladora* workers and their households needs to be understood in terms of the specific identities that could have been affected: as a male or a female Mexican, and as a *maquiladora* line worker. In effect, looking at the cultural origins of gender relationships, male and female *maquiladora* workers, as men and women, are "entitled to be perceived" according to the norms specific to the Mexican culture and history, which differ from other countries' social entitlements. The richness and depth of the Mexican identity is the result of centuries of ancient indigenous beliefs and customs together with a relatively recent imposition of Christian values by the *conquistadores* from 1519 onward. Published in 1950, the book from the 1990 Nobel Prize winner for Literature, Octavio Paz, entitled *"El laberinto de la soledad,"* is certainly the best description of the Mexican identity. As a Mexican citizen spending most of his life in foreign countries, Paz gives an objective and poetic view of Mexican behavioral norms and ideals coming from its unique history, which was a result of conflicts between various identities. Consequently, the social norms ruling Mexican behaviors are the result of this unique culture.

As Mexicans, the *maquiladora* workers inherit a particularly rich culture, which has been affected by the North American economic integration. The aim of this chapter is to offer a description of the post-NAFTA changes in social entitlements of male and female *maquiladora* workers. The first section looks briefly at the ancient gender ideals in the cultural makings of Mexican society. The second section then investigates the effect of these economic shocks on the Mexican identity by looking at the magnitude of changes in the social entitlements of the population resulting from the economic integration with the United States, and the 1994 shock in the Mexican currency, over the period 1994–2007. Various aspects of social entitlements are described, including population movements across the border, homicides, and economic growth in the border states. At the national level, descriptive statistics further describe changes in health indicators such as fertility rates, mortality rates, causes of mortality, and the final changes in the structure of the Mexican household. As *maquiladora* line workers, whether male or female, the same culture leads them to be entitled to different practices in the labor market. The third section of this chapter then looks at the gender identity inside the Mexican labor market in terms of changes in labor participation, changes in unemployment, changes in the gender composition by industry, and changes in employment practices. Finally, the last section confronts these cultural ideals with the current perception of gender ideals within *maquiladora* households revealed through a household survey.

5.1. Cultural Identity and Historical Ideals

In 2007, about 80 percent of the Mexican population considered themselves to be Catholics, against 88 percent in 2000, the remaining part of the population considered themselves to be Protestants, Evangelists, members of other religions, or atheists. In 2005, around 80 percent of the Mexican households surveyed stated that the norm is that the man is head of the household.[1] Combined with ancient indigenous beliefs, traditional religious beliefs lead to a hierarchal structure where the head of an institution is likely to be a male figure. The gender norms within the household are also present in the labor market. To understand the Mexican models of femininity and masculinity, one needs to dig deeper into the history of male and female ideals. This section therefore draws on Koch (2006), who wrote a detailed historical account of the foundations of the Mexican culture as a struggle between the *conquistadores* and the Aztecs.

When the first Spaniards arrived in the peninsula of Yucatan (south of Mexico) in 1519, the Aztec civilization was at its height. As for the Mayans, they were living on the Yucatan peninsula and were considered by the Spaniards as an advanced civilization, given their elaborate clothing and living conditions. However, at the time of the Spanish conquest, the Aztecs were the dominant indigenous community throughout Mexico, under the leadership of King Montezuma. One of the major gods in Aztec beliefs was Huitzilopochtli. Huitzilopochtli was a dark lord, thirsty for the sacrificial offerings[2] of living human hearts. Having been led to the Promised Land by Huitzilopochtli, the Aztecs considered themselves to be the chosen people. Before being called Aztecs, this core civilization experienced several name changes. These chosen people first appeared as the Chichimecs, and were then called Mexica, according to their leading priest who spoke the words of Huitzilopochtli. The Mexica were ferocious and fearless warriors. It was said that one of their warriors was worth several men on the battlefield. The prophecy of the Mexica was that they would settle at a certain place where an eagle was seen eating a snake while sitting on a cactus that had grown on a rock in the middle of a lake—the symbol of the current Mexican flag. Throughout the centuries from AD 900, Huitzilopochtli led the Mexica in a journey from one region to another, learning to survive with little means and practicing agrarian and building techniques picked up from the already decadent civilizations such as the Tolpecs. The Mexica, guided by Huitzilopochtli, saw the realization of the prophecy in the valley of Mexico, where they set up the city of Tenochtitlan, which was a living representation of the lost city of Atlantis, according to the Spaniards. The Spaniards also called it the "Venice of the New World," since it was established on an island in the middle of Lake Texcoco with numerous floating gardens and stone bridges. The conflict between the

Spaniards and the Aztecs was perceived to be the fight of the "Age of Iron" (e.g., guns) versus the "Age of Stone" at its best. The Aztecs' agrarian and architectural skills were state of the art, and the beauty of the city truly impressed the Spaniards. However, this city was burned down at the fall of the Aztec civilization in 1521 by the troops of Hernán Cortes, and upon its ashes were raised the foundations of today's Mexico City. Today, only a small part of the lake remains, as it has dried out over the centuries.

As part of his human representation, the god Huitzilopochtli was born from an immaculate conception between Coatlicue, the earth goddess, and Tezcatlipoca, a dark lord who, like his son, required a steady diet of live beating hearts. Huitzilopochtli was born under peculiar circumstances. Huitzilopochtli's brothers and sisters included sister moon and her brothers, the stars. Out of jealousy toward an illegitimate child, sister moon planned on killing her mother Coatlicue by enrolling her brothers, the stars, in her betrayal. However, one brother among the brother stars warned Coatlicue of her impending death. As she became frightened, the unborn Huitzilopochtli told his mother to rely on him to protect her. As sister moon started to attack her mother with her brothers the stars, the earth goddess Coatlicue was seized by the pain of birth pangs. Huitzilopochtli was then born as a full-fledged warrior with the strength of a god. He killed his sister the moon and sent her face up to heaven for eternity, thus saving his mother from certain death. Sister moon went into the space of the darkest nights, and human sacrifices were then made in order to make sure that the divine sun—made of three major gods, where all the souls of warriors killed on the battlefield would be as bright as the sunshine—would rise every morning. At the arrival of the Spaniards on the Aztecs' coast, they were welcomed by Aztec lords with, among other things, a golden plate representing the divine sun and a silver plate representing sister moon. In most Mexican households today, there is a representation of the sun next to a representation of the moon, which is a reminder of this duality between dark and light, the betrayal by sister moon and the heavenly comfort of the sun.

Huitzilopochtli's father Tezcatlipoca was also a god of dark power whose historical particularity was that he had chased out of the Mexican land the gentle god Quetzalcoatl. Quetzalcoatl was a king who preached the abolition of human sacrifices and advocated living with respect for one another. He dedicated himself to a life of purity, celibacy, and prayers for the good of his people. His time as king was a time of peace and serenity that was soon disturbed by the evil Tezcaltlipoca. By haunting Quetzalcoatl's dreams and pushing him to commit sins, Tezcatlipoca managed to evict Quetzalcoatl from the known world. When Quetzalcoatl left, he had time to preach to other tribes on his way to the eastern coast of Mexico. From there, he headed off toward the rising sun across the Atlantic Ocean, beyond the known world, making a promise to his people that he would return. Thanks to the similarities with some Christian beliefs and to the arrival of the Spaniards from the eastern coast, Hernán Cortes, whose conquest of Mexico began in 1519, used this unforgotten legend to fool the Aztecs by portraying the Spanish as gods or as Quetzalcoatl's disciples. Hernán Cortes led the Spaniards to the conquest of Tenochtitlan thanks to alliances with other Aztec rivals, which led to numerous deaths on both sides, and thanks also to his translator-lover Dona Marina, a beautiful native who was Christianized by the Spaniards. Her considerable role in the Spanish conquest, translating and giving valuable cultural insights, made her invaluable in the Spanish eyes, whereas she was considered as a traitor by the indigenous population.

In the light of these historical beliefs, it seems that a powerful image of gender roles emerges from both traditions. With the enforcement of Christian beliefs upon the indigenous traditions, a final and major Mexican symbol needed to be introduced: the symbol of the Virgin of Guadalupe. The Virgin appeared to an Indian during the Spanish conquest at the same place where the earth and fertility goddess was worshiped in pre-Hispanic times. Nowadays, the Virgin of Guadalupe remains a dominant emblem of Mexican society, linking together family, politics, and religion, the colonial past and independent present, and indigenous and Spanish descendents (Wolf 1958). In indigenous belief, the female figure can be associated with several characters: the earth goddess who was saved by her son Huitzilopochtli, the sister moon who tried to kill her mother earth, or the image of Dona Marina, who betrayed her Aztec origins by helping the *conquistadores*. In Christian belief, the female figure was then transposed to a holy mother, the Virgin Mary, who had also experienced an immaculate conception, as did the Goddess Earth. The Virgin of Guadalupe unites the religious sensibilities of the Mediterranean and Mesoamerica, as both of these had fostered ancient cults of feminine divinities (Paz 1985, 366). Thus, the traditional image of woman is that of a perfect virgin mother and provider of emotional warmth. However, from the indigenous aspect of the tradition, this image also includes a threat of betrayal.

On the other hand, the male image given by the Christian ideal is one of a God-made man, which is similar to the Aztec ideal. In the Aztec tradition, the male figure can be associated with the son Huitzilopochtli, savior of his mother, and his father Tezcatlipoca, the dark lord, but also with the God Sun, god of the warriors, provider of comfort and light. In the Christian tradition, the male model is represented by Jesus Christ, son of God and savior of all human souls (Aquinas 2006). Published in 1950, Paz's account of the Mexican man states that "the manly ideal consists in an open and aggressive fondness for combat, whereas we emphasize defensiveness, the readiness to repel any attack. The Mexican *macho*—the male—is a hermetic being, closed up in himself, capable of guarding both himself and whatever has been confided to him" (Paz 1985, 31). In Paz' analysis of the relationship between the United States and Mexico, the manly ideal is close to his view of capitalism that exalts the behavioral patterns traditionally perceived as virile, such aggressiveness and combativeness.

The Mexican culture reflects the universal dichotomy of femininity versus masculinity with its a unique set of norms of behavior toward gender roles. Within households, by displaying representations of the sun next to the moon, and the Virgin of Guadalupe next to a crucified Jesus, it restores and sustains an equilibrium between specific icons coming from mixed traditional values. When external shocks hit this equilibrium of femininity and masculinity, where equilibrium is understood as a point at which there is no tendency for change, this may create conflicts of cultural identities that go beyond gender issues.

> The Mexican, heir to the great pre-Columbian religions based on nature, is a good deal more pagan than the Spaniard, and does not condemn the natural world. Sexual love is not tinged with grief and horror in Mexico as it is in Spain. Instincts themselves are not dangerous; the danger lies in any personal, individual expression of them...Modesty results from shame at one's own or another's nakedness, and with us it is an almost physical reflex. Nothing could be further from this attitude than that fear of the body which is characteristic of North American life. We are not afraid or ashamed of our bodies; we accept them as completely natural and we live physically with considerable *gusto*. It is the opposite of Puritanism. The body exists, and gives weight and shape to our existence. It causes us pain and it gives us pleasure; it is not a suit of clothes we are in the habit of wearing, not something apart from us: we *are* our bodies. (Paz 1985, 35)

Under NAFTA rules, however, illegal Mexican "bodies" must stop at the border. In reality, the Mexican identity and associated norms of behavior have been spreading across the United States among the Mexican immigrants, legal and illegal. Similarly, US identity has been spreading across Mexico, since NAFTA implementation, through the flows of goods and services, legal and illegal. The border region is at the heart of this fusion. The economic integration of Mexico with North America may have led to a violent clash of identities between Mexican and North American ideals at the border, where there is equal inhibition of both Mexican ancient values and North American puritan values.

5.2. External Shocks on Mexican Identity

All economies experience welfare changes when making the transition from developing countries to developed countries.

However, the global contexts in which these changes take place affect in greatly different ways the welfare outcome in each country.[3] Since NAFTA implementation in 1994, real Mexican GDP increased by more than 35 percent, but the economic integration of the Mexican economy with North American economies came at a high price, since the Mexican CPI was multiplied by ten over the same period, as shown in figure 4.1 in chapter 4. After the debt crisis and the slow growth of the Mexican economy in the 1980s, the 1990s and NAFTA implementation came with the hope of new economic opportunities for the Mexican population. These hopes were suddenly slashed by the peso devaluation in December 1994 leading to a contraction of GDP and rising inflation in 1995. However, in 1996, the Mexican economy started to recover. As shown in table 5.1, prior to the implementation of NAFTA and devaluation of the peso, the Mexican GDP increased by only 1.6 percent from 1990 to 1995. After these events, the GDP grew by an impressive 25.3 percent from 1995 to 2000, which helped to support a growing middle class, before being slowed down by the US recession in 2001, with GDP increasing by only 6.5 percent from 2000 to 2005.

Economic growth triggered by an event such as currency devaluation, through a rise in the demand for exports, also weakens the purchasing power of the poorest and youngest sections of the population. In difficult times, these people seek the highest possible income to improve, or at least sustain, their purchasing power and associated living standards in export-led industries or abroad. Home of the export-led industries, including the *maquiladora* industry, the border region with the United States reflects this labor mobility with the population movements it engenders. Table 5.1 shows the population and GDP growth in the Mexican border states and the population movements across the Mexico-US border over the 1990–2005 period. The GDP growth rates mentioned above, at the national level, are broken down according to the border states, namely, Baja California, Coahuila de Zaragoza, Chihuahua, Nuevo Leon, Sonora, and Tamaulipas. With a lower magnitude, the national trend follows the extreme economic changes happening in the border states. Baja California and Chihuahua

Table 5.1 Population movements in border states according to GDP growth (percentages)

	1990–1995	1995–2000	2000–2005	2005–2010	1990–2005
*Population growth (annual average over each period)**					
National	2.1	1.6	1	1.8	1.8
Baja California	4.3	3.8	2.4	5	3.9
Coahuila de Zaragoza	1.7	1.3	1.6	1.7	1.5
Chihuahua	2.4	2.1	1.1	1	1.8
Nuevo León	2.4	1.8	1.6	2.1	2
Sonora	2.4	1.4	1.4	2.1	1.8
Tamaulipas	2.1	2	1.7	1.6	1.9
GDP growth (over each period)					
Real GDP National	1.6	25.3	6.5	8.1	39.7
Nominal GDP National	27.5	21.8	43.7	31.1	129.5
**Baja California	1.1	51.8	10.8	-1.9	70.1
Coahuila de Zaragoza	2.7	37.2	17.5	-3.1	65.6
Chihuahua	-0.9	48.8	11.5	0.1	64.3
Nuevo León	-1.3	39.4	16.7	5.1	60.6
Sonora	4.7	30.5	10.9	7.6	51.6
Tamaulipas	1.5	37.5	18.5	0.3	65.4

	1992	1995	1997	2000	2002
*Migration to USA *** 15–24 years*	48.3	49.6	48.6	68.6	41.5
25–49 years	38.9	35.7	41.8	22.2	44.8
Migration from USA 15–24 years	36.9	33.8	33.9	55.1	43.4
25–49 years	45.7	42.8	53.9	35.4	36.7

Source: Author's calculations from INEGI data (2011), Censo de Población y Vivienda, and Sistema de Cuentas Nacionales de México.

Notes: *The Mexican population totaled 103 million in 2005 with a 51.4 percent female population, against 97.5 million in 2000 with a 51.2 percent female population.

*The GDP data concerning border states is in real terms and is available from 1993 to 2009 only; the period 1990–1995 therefore covers the period 1993–1995 and the period 2005–2010 covers the period 2005–2009.

***Migration data relates to growth rates of migrants in each age group to and from the United States for the past five years of each year stated. w experienced twice the growth rate of the national level from 1995 to 2000, with 51.8 percent and 48.8 percent, respectively. Five years later, for the period from 2005 to 2010, these rates become –1.9 percent for Baja California, and 0.1 percent for Chihuahua. The other four border states experienced a similar trend, apart from Nuevo Leon and Sonora, which are closer to the national trend.

In terms of population growth over the period from 1990 to 2005, the annual population growth in the border states is around 2 percent, which is similar to the national trend. Only Baja California experienced twice the population growth rates of the national level. Finally, data on migration flows to and from the United States, based on population surveys, reflects the trend, rather than the magnitude, of both legal and illegal migration. The migration rate shown relates to a growth rate in migration for the five years prior to the year stated. For example, the 2000 figure shows that migration across the border has substantially increased over the period 1995–2000 for the 15–24-year group, and that it has substantially decreased for the older age group of 25–49 years compared with other years over the available period from 1990 to 2002. Since NAFTA implementation, there has been an acceleration of trade in goods and services only. Coupled with peso devaluation and the resulting lower purchasing power of the peso, this means that the cheap Mexican labor force continues to remain frustrated at the border, attempting illegal migration by all means, or working in export-led industries or in the informal sector.

An important element of Mexican social entitlements in the border states is that this region is a geographical center connecting two conflicting systems of human functionings, which leads to a clash of identities. Paz (1985), for example, explained how, even in pre-Columbian times, the current Mexican territory was the home of settled farmers while the current US territory was home of hunters and nomads. This border region is a fascinating case study for global integration because it is witnessing a fusion of two cultural systems of behavioral norms with two different levels of economic development. Table 5.2 shows the growth rates in accidental or violent homicides from

Table 5.2   Growth rates in accidental or violent homicides in border states (percentages)

	1990–1995	1995–2000	2000–2005	2005–2009
Total				
National	−3.4	−8.4	1.9	26.9
Baja California	15.1	11.2	4.6	49.9
Coahuila de Zaragoza	−15.2	−6.4	15.7	23.9
Chihuahua	3.3	−10.4	3.5	131.1
Nuevo León	0.7	0.3	12.1	19.4
Sonora	2.2	−4.6	11.0	30.4
Tamaulipas	−4.2	3.3	3.2	2.3
Men				
National	−2.2	−9.6	−0.2	31.3
Baja California	20.0	8.8	1.5	57.9
Coahuila de Zaragoza	−10.2	−9.6	13.5	25.6
Chihuahua	9.2	−10.4	−0.4	156.9
Nuevo León	0.9	−1.3	6.5	21.4
Sonora	9.8	−10.3	13.8	28.5
Tamaulipas	−1.2	2.4	0.6	1.2
Women				
National	−7.0	−3.1	10.4	10.5
Baja California	−4.5	24.6	18.0	17.9
Coahuila de Zaragoza	−30.8	7.2	25.8	17.9
Chihuahua	−12.8	−10.8	22.5	28.2
Nuevo León	0	7.8	15.8	13.7
Sonora	−25.4	26.4	0.7	37.4
Tamaulipas	−13.9	8.3	13.0	7.4

Source: Author's calculations from INEGI data (2011), Estadísticas de mortalidad.

1990 to 2009 in the border states, including death linked to criminality such as drug traffic, illegal migration, and kidnapping. Looking at table 5.2, the national trend of homicides from 1990 to 2000 is moving downward, while from 2000, and especially from 2005, there is an impressive resurgence of homicides, with an increase of 27 percent from 2005 to 2009. Chihuahua is one of the main contributors of that sudden increase, as male homicides here have increased by 157 percent between 2005 and 2009, which is a sign of the increase in violence linked to the drug war. Since 1990, Baja California has become a riskier border state to live in given its growing trend of both male and female homicides in each period. Finally, looking at the gender characteristic in homicide rates, there are mostly positive growth rates for female homicides since 1995, while there are negative growth rates visible for male homicides up to 2005. In effect, while all female homicides declined from 1990 to 1995 across the border states, the victims of criminal behavior have crossed the boundary of gender since 1995. An example of this rising trend is the well-known phenomenon involving violent homicides of female *maquiladora* workers since 1993,[4] particularly in Ciudad Juarez (Chihuahua), which is argued to be "the global economy's new killing fields" (Bowden 2010).

Painting a picture of the changes in social entitlements of the Mexican identity in the border states will not be complete without looking at the changes in welfare standards at the national level. Human beings are known to adapt to their environment, but the process of adaptation is made according to certain standards, such as the standards of consumption behavior. Economic changes that suddenly bring about new standards may not be absorbed by long-established human functionings built according to local environmental standards. The speed at which economic changes occur or their magnitude may exacerbate their impact on a given population, which may be in terms of income distribution, demographic changes, or metabolic changes. Table 5.3 displays a number of figures on a range of health aspects of the Mexican population from 1990 to 2005. Looking at the mortality rates and causes of mortality, there is a 33 percent increase in mortality from 1990 to 2005. Looking deeper into that issue, the rise in mortality related to diabetes is the most striking of all causes of mortality. In 2006, diabetes became the second cause of mortality for both men and women, while the first cause was liver diseases, for men, and malignant tumors, for women. Mortality caused by diabetes increased by 173 percent for men and by 139 percent for women from 1990 to 2006. The second place for the most important cause of increase in mortality is shared between liver diseases and malignant tumors, depending on the gender of the population affected, which increased by around 50 percent over the same period. These figures may suggest that Mexicans have made many changes in their nutritional habits, and that their physical functionings are not able to adapt to the new standards in their diet. Diabetes and liver diseases have become the main causes of mortality in Mexico for both men and women.

The period associated with high GDP growth rates from 1995 to 2000 is also associated with the concentration of all human resources into the labor market. This led to the economic independence of female household members, which became associated with increased divorce rates and decreased fertility rates. From 1990 to 2005, the population grew by 27 percent, but on a constant decreasing trend, with the rate of population growth being 12 percent from 1990 to 1995, 6.9 percent from 1995 to 2000, and 5.9 percent from 2000 to 2005. The fertility rate fell from 3.4 children in 1990 to 2.2 children in 2005, with the variations in the percentage changes being –14 percent from 1990 to 1995, –3 percent from 1995 to 2000, and –21 percent from 2000 to 2005. In line with the changes in fertility rates, education levels increased (not displayed in the table 5.3), with, for example, a rise in university attendance for women aged between 20 and 24 years. This increase was from 13.8 percent for this age group in 1990 to 19.6 percent in 2005—this increasing trend was also visible for men, with 17.9 percent and 22.2 percent in 1990 and 2005, respectively. The changes in the social organization of Mexican society thus went through women empowerment, increased female participation in the labor market, and better access to education for both men and women.

The ideals of gender roles in the Mexican culture show that the household structure remains strongly attached to the nuclear family model and to the family values related to this household structure. Over the past 50 years, nuclear households formed a constant 98 percent of all households. However, the period from 1995 to 2006 saw substantial changes in diverse aspects of the household structure. Table 5.4 presents data on the growth rate of male-headed households, female-headed households, divorces, and marriages over the period from 1990 to 2005. The Mexican household structure seems to have changed over the past two decades. Prior to NAFTA's implementation in January 1994 or prior to the December 1994 peso devaluation, the growth rate for divorces showed a fall of 14 percent

Table 5.3 Growth rate in different aspects of health (percentages)

Health aspects	1990–1995	1995–2000	2000–2005	2005–2010	1990–2005
Population growth	12.1	6.9	5.9	8.8	27.1
Fertility rate*	-14.7	-3.4	-21.4	-4.5	-35.3
Mortality rate					
Total	1.8	1.7	13.1	14.0	33.5
Men	1.4	0.8	11.8	15.7	32.2
Women	2.7	2.9	14.9	11.9	35.9
Causes of mortality: **					
Heart diseases					
Men (4th cause)	17.5	4.9	19.0	n.a.	46.8
Women (3rd cause)	15.4	1.9	4.9	n.a.	23.5
Malignant tumors					
Men (5th cause)	16.7	12.8	15.5	n.a.	52.2
Women (1st cause)	14.4	12.1	15.8	n.a.	48.5
Diabetes					
Men (2nd cause)	33.7	36.2	49.7	n.a.	172.7
Women (2nd cause)	30.2	35.1	35.7	n.a.	138.8
Liver diseases					
Men (1st cause)	13.8	23.9	0.6	n.a.	41.8
Women (4th cause)	20.2	21.5	2.6	n.a.	49.9

Source: Author's calculations from INEGI (2011), Censo de Población y Vivienda and Estadísticas de Mortalidad.
Notes: * Fertility rates were 3.4 children per woman in 1990 and 2.2 in 2005.
** The rating of causes is made according to the 2005 classification.

Table 5.4 Growth rate in different aspects of household structure (percentages)

Aspects of household structure	1990–1995	1995–2000	2000–2005	2005–2008	1990–2005
Nuclear households*	19.9 (1990–2000)		5.1	n.a.	26.0
Males as head of households	15.3 (1990–2000)		2.3	n.a.	18.0
Females as head of households	48.1 (1990–2000)		18.3	n.a.	75.2
Divorces	−14.4	35.6	42.3	16.6	65.2
Marriages	2.5	7.5	−15.8	−1.1	−7.2

Source: Author's calculations from INEGI (2011), Censo de Población y Vivienda.
Note: * Nuclear families are constant at around 98 percent of all households from 1950 to 2005.

over the 1990–1995 period, which was followed by an impressive 35 percent increase from 1995 to 2000 and a 46 percent increase from 2000 to 2005. Meanwhile, the growth rate of marriages rose slowly from 1990 to 2000 and started to decline from 2000 to 2005, with a growth rate of −12 percent during this period. Consequently, the growth rate of single-headed households saw a dramatic increase, especially female-headed households, with an increase of 75 percent from 1990 to 2005. Thus, these figures suggest a change happening in the social organization of Mexican society, which is confirmed by the changes in GDP growth, population movements, and public health over the same period.

5.3. Gender Identity in the Labor Market

The cultural standards of the Mexican identity are integrated into the social norms and practices. The typical features of such cultural background include a tendency for paternalistic behavior on the part of all institutions, from the government, and the firm, to the household. Sen has argued in the past that "the issue of paternalism, when it does arise, must relate to the rejection of the person's self-evaluation" (1987, 82). Paternalism includes a degree of hierarchical organization whereby one identity is dominant over others, and this is then followed by a perceived order from the most valued to the less valued identity. The self-evaluation of any identity will be biased according to this hierarchical organization whenever paternalism is the norm of fairness to which people refer to as an optimality point in social interactions.

In the context of the gender identity in the Mexican labor market, some commentators put forward the argument of a segmented labor market in which the discrimination of employers against women in some sectors would create an oversupply of women in other sectors, thus pushing down wages in those sectors (Katz and Correia 2001). From 1987 to 1995 there were substantial changes in the gender composition of the labor force in Mexico, notably due to the economic liberalization and trade expansion of the Mexican economy. In terms of trade sectors, there has been an increase in female employment, from 31.2 percent to 35.4 percent of the total labor force in the export-oriented sector, and from 61 percent to 72.4 percent for the nontradable sector, in the period from 1987 to 1993 (Ghiara 1999). Table 5.5 shows the labor force participation, unemployment, and informal employment of the Mexican working-age population from 2000 to 2007. It can be mentioned that female participation in the labor force

Table 5.5   Employment and unemployment rates by gender (2000–2007)

Years	Labor force participation (% of working age population)		Unemployment (% of labor force)		Informal employment (% of labor force)	
	Men	Women	Men	Women	Men	Women
2000	80.2	36.9	n.a.	n.a.	n.a.	n.a.
2001	79.7	36.7	n.a.	n.a.	n.a.	n.a.
2002	79	36.9	n.a.	n.a.	n.a.	n.a.
2003	78.8	37.3	2.9	4.3	n.a.	n.a.
2004	78.5	38.9	3.3	5.1	n.a.	n.a.
2005	78.2	40.2	3.4	4	27.5	29.3
2006	78.6	41.3	3.4	3.9	26.5	28.4
2007	78.4	41.5	3.5	4.1	26.5	27.4

Source: INEGI (2011).

has persistently increased over the past decade, from 36.9 percent of the working-age population in 2000 to 41.5 percent in 2007. The female labor force, skilled and unskilled, entering the labor market has not affected all sectors of the economy evenly. The impact of trade liberalization on the participation of the female labor force has been discussed by Ghiara (1999) in the context of Mexico. Looking at the impact of trade liberalization on female wages across sectors, Ghiara's conclusions regarding the industries employing a higher-skilled female labor force are relevant in the context of the female-led industrialization school. According to Ghiara (1999), female-led industrialization theory states that as the demand for female labor increases, women's bargaining power in asking for higher wages and better working conditions also rises. The sectors concerned, in the context of Mexico, are the education and health sectors or the services sector, where most of the educated female labor force is concentrated. This theoretical perspective can also be related to bargaining and efficiency wage models arguing that there is a positive relationship between wages and employment.[5]

The movements of women labor across industries are linked to the increase in the overall female labor force, as shown in table 5.6 and as suggested by the changes in household structure in the previous section. Table 5.6 shows the growth rates in labor force participation by gender and per industry from 1998

Table 5.6 Growth rates in labor force composition per industry (1998–2003)

Industry	Men	Women	Total
Agriculture and fishing	+15.5	+169	+21.1
Mining	−3.8	+25.3	−1.4
Manufacturing	−11.2	−4	−8.8
Electricity and water	+10.5	+27.3	+13
Building	−4.7	+15.5	−3.7
Trade	+20	+24.4	+22
Transport and communication	−31.9	−17.3	−30.2
Financial services	−9	+4.8	−3.4
Social services	+3.1	+18.7	+9.3
Total	10.3	114.7	15.4

Source: Author's calculations from INEGI (2011).

to 2003. The decline of the labor force is more pronounced for men compared to women in industries such as manufacturing, transport, and communication. The participation of women in manufacturing industries fell by 4 percent from 1998 to 2003, as against a corresponding figure of –11.2 percent for men. The most striking change in female labor force participation is in agriculture and fishing, which showed a rise of 169 percent over the period from 1998 to 2003. The demand for a low-skilled labor force in diverse sectors of this industry may be one reason for attracting low-skilled first-time entrants to the labor market.

Flexible and low-skilled female labor force can be drawn directly from the household economy or from the informal sector where they were previously employed. The data concerning the informal sector in table 5.5 does not cover a large time span because of the challenges involved in collecting official and accurate statistics on the informal sector. However, the trend shows a decrease in employment in the informal sector for both men and women from 2005 to 2007. It can only be assumed that this decreasing trend started in the previous years. It remains to date that informal employment accounts for more than 25 percent of the labor force for both men and women. Considering the informal sector as a "reserve army" of discouraged workers or low-skilled potential workers for the *maquiladora* industry would mean a lower bargaining power for workers within the industry. The informal sector is the main source of income for the poor and vulnerable; however, their reserve wage is lower than the money wage offered in the formal labor market. Unskilled poor women looking for jobs in urban areas rely in part on factory-type of work in factories or through subcontracted work at home, or on paid domestic work.[6]

Considering jobs in the informal sector as a form of underemployment that provides work at what is, supposedly, the reservation wage is an important issue, since the literature related to the relationship between real wages and unemployment becomes of great relevance here. The informal sector is characterized by the fact that workers are not unionized and are not entitled to minimum wages or subject to taxation. The existence and spread of the informal sector is a significant element for the process of wage determination within the formal sector. For example, empirical evidence from Taiwan has highlighted that there is a significant negative relationship between local unemployment and the level of wages (Rodgers and Nataraj 1999), as Blanchflower and Oswald (1994) had already argued. For official institutions, such as the Mexican Official Institution for Statistics (INEGI 2011) and the International Labor Organization (ILO), the informal sector is considered to be a form of employment with different characteristics. The official Mexican definition of the informal economy published by ILO states that these are unincorporated enterprises that are owned by households, producing all or parts of their goods or services for sale, and having 16 or less persons engaged (for manufacturing) or six or less persons engaged (for other branches of economic activity), which excludes paid domestic workers and workers in quasi-corporations (KILM 2005). INEGI considers the informal sector as self-employment of ten or less persons, which refers to all market activities produced from household resources independent from those households, and without being a registered firm per se (INEGI 2011). The distinction between the informal sector and the formal sector is crucial in understanding the reserve army of workers employed in the informal sector, remuneration and working conditions being the key elements of comparison between the two.

Depending on the definitions of unemployment and the informal sector used by official institutions, there will be different policy implications in terms of setting policy standards for tackling the informal sector, either coercively or by supporting self-employment. Over the past 20 years, the informal sector represents a constant 30 percent of total employment for each gender—which suggests that there is gender equality in access to the informal sector, while male participation is double the female participation in formal employment (KILM 2005). According to the INEGI[7] definitions used in their population surveys, and as mentioned above, the informal sector is accounted in employment figures as self-employment, which takes place in a household enterprise that is not registered with the state and comprising less than ten people. Since part of the informal sector, defined as illegal work with nontaxable income, is counted as self-employment in official statistics, it explains

why the rate of unemployment at the national level has continued to remain around 5 percent over the past decade, according to the official institute for national statistics (INEGI 2011).
Individuals working in the informal sector are underemployed
and/or underpaid in terms of the actual work performed; given
its illegal nature, no protection or rights are ensured for the
worker. Underemployment can also be defined as the misuse
of one's skills or working less hours than one would have liked
to, which is quite challenging to capture with official statistics.
Also, a higher employment share in the informal sector could
weaken the bargaining position of formal sector workers, as
there will be a large pool of workers to replace them, which is
consistent with the literature on the informal sector.[8] The form
of underemployment in the informal sector therefore puts more
pressure on the workers' economic entitlements than does the
form of underemployment in the formal sector. INEGI defines
underemployment as the situation where working-age population employed in the formal sector look for employment for
additional working hours. Their statistics suggest that, in 2007,
seven out of a hundred people in formal employment were
underemployed.

Just as the institutional boundary between underemployment, formal employment, and informal employment is difficult to draw, so is the perception of gender identity and
associated norms or behavior between the formal and informal
sectors, including the household economy. In the formal labor
market, gender imbalances impact on the balance of power over
the use of resources and lead to an unequal access to income
for female workers. Ghiara's empirical findings mentioned
above suggest that the male-female wage differential in low-
skilled industries in Mexico, such as the *maquiladora* industry,
seems to be consistent with the so-called female marginaliza-
tion theory. According to Ghiara (1999), this theory argues
that as the demand for female labor rises, wages do not follow
the same pattern as applicable to men's wages when male labor
demand rises. An increase in the demand for female labor gen-
erates smaller increases in female wages than is applicable to
male wages, and thus the gap between male and female wages
widens further. The mechanisms involved here are based on

the effects of rising demand for the two different labor supply functions of men and of women. From an economist's point of view, this is a form of the monopsony approach proposed by Robinson (1969),[9] whereby one buyer, the firm, faces a large number of sellers, male and female workers, with an upward sloping curve. In such a case, a monopolistic firm can find it more profitable to discriminate against a group of workers with a lower elasticity of supply by offering a lower wage to this group than to a group with a higher elasticity of supply. Given the importance of informal employment for the labor force, and given the importance of other export industries in Mexico, the main reason for not using the monopsony hypothesis is that the *maquiladora* industry is not the sole buyer facing a large number of unskilled workers. Industries, other than the *maquiladora* type of industry, contribute to national output and are ready to employ a large number of unskilled workers, male or female.

The gender empowerment measure (GEM) of the Human Development Index captures the gender inequalities occurring in a given country in three different areas: political participation, measured by women's and men's percentage shares of parliamentary seats; economic participation, measured by their shares of positions as legislators, managers, professional and technical positions; and power over economic resources, measured by women's and men's estimated earned income. In 2005, Mexico was ranked 38th out of more than 170 countries (UNDP 2005). However, looking at this index closely, it can be seen that Mexico's rank drops to 127 out of the same 170 countries when its ratio of female earned income to male earned income is considered, which is 0.38. This seems to support the argument that the gender income gap within the household, whereby men earn a larger share of the household income, is translated into gender disparities in the labor market, whereby men are entitled to a higher wage than women in the same job position. The threshold for the socially acceptable fair wage per male effort is placed higher than the fair wage per female effort. The social norms in place warrant that the notion of fairness places a higher monetary value on male effort than on female effort. If the effort spent in the labor market is rewarded by wages and the effort spent in the household is not, the opportunity cost of the effort displayed in the labor market in terms of the effort spent in the household is socially perceived to have a higher monetary value for men than for women. The practice of gender norms in Mexican society based on strong cultural and religious influences affect social interactions within the household and inside the labor market, and strongly influences the perceived market value of social identities.

5.4. Gender Identity within Maquiladora Households

Models of femininity and masculinity in the culture and traditions of a population contribute to people's imagination of an idealistic view of male and female models of behavior. At the household level, norms then lead to individuals' expectations about how the other gender should behave, and how resources should be allocated according to a gender and cultural benchmark of optimality. Social norms are a matter of perception about how people should behave according to a personal idealistic vision attached to an identity, which here is in terms of gender and cultural identity. Given the cultural and gender ideals in Mexico, the aim of this household survey is to reveal whether there are differences of perception of optimal resource allocation between individuals, and the factors affecting individuals' sense of optimality. Individuals adapt their preferences according to a benchmark of optimality, which is influenced by identity-related ideals. In effect, whether individuals have enough access to a commodity or not, or whether they have enough access to a potential functioning, is decided not simply in comparison with others, but in comparison with the ideals that others represent. Within the household this is especially visible because norms and ideals constrain the interdependence of opportunity sets within a limited geographical space. For example, the use of a car can be a collective occurrence but driving a car is an individual action. Assuming a household owns a car, its collective use will depend on who the ideal driver will be. If norms are attached to the identity of this ideal driver, the collective use makes it reliable on this identity. Over time, preferences adapt to these ideals as each individual gets used to the identity of the ideal driver.

By focusing on the two heads of the household, husband and wife, in *maquiladora* households of the major northern city of Monterrey, this empirical analysis carried out in 2006 investigates the cross-perceptions of gender norms among household members and its impact on resource allocation within the household, as perceived by each household member. Thus, controlling for actual allocation of resources, gender, income, job tenure, education, age, and household structure, the difference in factors affecting the optimal allocation for the households by husbands and wives is expected to reflect cultural and gender norms in the population sample. Data on individual characteristics collected include individual salary, job, age, education, and salary in the previous job, if any. These personal characteristics serve as control variables or objective variables, while the vectors described below constitute the subjective variables. Clark (2003) investigated the perception of well-being among the poor and found that the people surveyed referred to the following items associated to a "good life": jobs, housing, education, income, family and friends, religion, health, food, good clothes, recreations and relaxation, safety, and economic security. Assuming that this is one of the definitions of a "good life," it is also replicated in the *maquiladora* household survey. In all, there are 20 items that describe all the major entitlements that can be expected for a good life, assuming that this is just one of the definitions being used. The items considered in this study include achieved functionings—for example, receiving affectionate gestures representing the functioning of being cherished— and commodities (goods, services, and public goods) as the items that the individual owns and/or makes use of. These items are related to consumption (food, clothes, medical care, education), leisure time (watching TV, leisure, spiritual life), housework (sewing and handcraft, cooking and home cleaning, repairs and driving), labor market (job, income, job benefits), transportation means (car, bus, and bicycle), and social interactions (giving/receiving affectionate/aggressive gestures). Similarly to Djebbari (2005) in her intra-household analysis of power distribution, the three categories, sewing and craft-making, cooking and home cleaning, repairs and driving, are included to add to the gender character of daily individual functionings within the household.

In the survey conducted, two questions were asked for each of the resources investigated. First, it was asked whether the respondent "has got or experiences the following," depending on the nature of the resources, as a good, a service, an activity, or a gesture. Taking the value to be 1, if yes, and 0, if no, the answers for all items represent the individual's endowment vector. The second question asked is whether the respondent is sufficiently entitled to the item, using the levels "not enough," "enough," and "a lot"—taking the values 0, 1, and 2, respectively. The answers for all items represent the individual's entitlement vector.[10] Finally, and most importantly for the purpose of this research, the individual is asked to repeat the exercise for each of the other household members, thus revealing their own perception of others' entitlements, whether they be the spouse, children, or any other relative living in the household. This last exercise gives a set of endowment and entitlement vectors describing how the individual perceives the household and vice versa. The small-scale household survey gives a population sample of ten control households and twenty *maquiladora* households. Ten out of the twenty *maquiladora* households include female workers and the remaining ten include male workers.

Cultural and gender differences can be revealed by looking at the optimality levels in endowments and entitlements, as perceived by husbands and wives. Table 5.7 presents the spouses' perception of their own actual allocation of resources. The higher the mean comprised between 0 and 1, the higher the sense of ownership of husbands and wives for all resources. The sense of ownership is rather higher for husbands, with a mean

Table 5.7   Endowment of resources by gender as perceived by the gender*

	Mean endowment	N	Std. deviation
Husband	.81	381	.396
Wife	.75	465	.433

Note: *The higher the mean (0<mean<1), the higher the sense of ownership for all resources.

of 0.81, than for wives, with a mean of 0.75. However, husbands were less likely to provide answers to questions posed, given that 381 answers were collected for husbands versus 465 answers for wives.

Table 5.8 displays the breakdown of optimality levels of husbands and wives by resource from the point of view of the spouse, and from the point of view of the household. The mean lies between 0 (not enough) and 2 (more than enough), with 1 representing enough. In other words, the closer the mean is to 1, the closer the perception is to the benchmark or ideal allocation represented by the level "enough." For example, looking at access to the labor market in the first set of columns and the "job" resource in particular, husbands are seen to be working more than they would ideally like, with a mean of 1.17, while wives are working less than they would ideally like, with a mean of 0.65. However, turning to the second set of columns in table 5.8, wives perceive that husbands work enough, with a mean of 0.97, while husbands perceive that wives are working less than what they ideally should, with a mean of 0.76. Both husbands and wives think that wives should have better access to the labor market. Wives' expectation of husbands' participation in the labor market is close to wives' optimal view of what husbands should do, while husbands think their labor market participation is higher that it should ideally be. It could be the case, for example, that husbands' labor market identity, as employer or employee, is taking over their household identity, as father, son, or husband, given that they perceive their leisure time to be less that they would ideally like, with a mean of 0.68. Considering both husbands and wives to be breadwinners, the "income" resource makes it clear that both husbands and wives agree that their individual financial contribution to the collective income is less than it should ideally be for the optimal well-being of the household.

Looking at the total means, husbands and wives display equal optimality levels of 0.76 and 0.77, respectively. Adaptive preferences according to gender and cultural ideals are likely to be responsible for this convergence. Husbands and wives tend to function together according to an optimality point of collective functioning. Individual behaviors adapt to this collective

Table 5.8 Perceived optimality between spouses

Resources	Gender	Spouses' optimal allocation as perceived by themselves			Spouses' optimal allocation as perceived by the other spouse		
		Mean*	N	Std. deviation	Mean*	N	Std. deviation
Consumption							
Food	Husband	.95	20	.394	1.11	19	.567
	Wife	1.04	25	.539	.80	22	.398
Clothes	Husband	1.00	19	.471	1.03	19	.716
	Wife	1.00	24	.659	.80	22	.480
Medical care	Husband	.79	19	.419	.86	18	.589
	Wife	.71	21	.644	.50	22	.463
Education	Husband	.84	19	.765	.95	19	.685
	Wife	.61	23	.722	.50	22	.598
Leisure time							
TV Hours	Husband	.53	19	.513	.82	19	.749
	Wife	1.00	23	.603	.75	22	.506
Leisure	Husband	.68	19	.582	.97	19	.539
	Wife	1.00	23	.674	.50	22	.512
Spiritual Life	Husband	.44	18	.616	.50	19	.687
	Wife	.70	23	.703	.33	20	.568
Housework							
Sewing/	Husband	.60	10	.699	.42	13	.703
Hand-craft	Wife	.27	22	.456	.47	18	.499
Cooking/	Husband	.56	16	.629	.39	19	.488
Cleaning	Wife	.80	25	.577	.86	21	.839
Repairs/	Husband	.82	17	.636	.82	17	.660
Driving	Wife	.44	18	.705	.35	17	.493
Labor Market							
Job	Husband	1.17	18	.514	.97	18	.629
	Wife	.65	20	.671	.76	21	.682
Income	Husband	.65	17	.493	.72	18	.548
	Wife	.55	20	.605	.40	21	.584
Job benefits	Husband	.71	17	.470	.78	18	.647
	Wife	.65	17	.702	.42	19	.449
Transport							
Car	Husband	.75	12	.754	.92	13	.838
	Wife	.78	9	.667	.79	12	.838
Bus	Husband	.85	13	.555	1.07	14	.829
	Wife	1.00	18	.686	1.03	17	.695
Bicycle	Husband	.60	5	.548	.19	8	.372
	Wife	.00	4	.000	.11	9	.333

Continued

Table 5.8 Continued

Resources	Gender	Spouses' optimal allocation as perceived by themselves			Spouses' optimal allocation as perceived by the other spouse		
		Mean*	N	Std. deviation	Mean*	N	Std. deviation
Social interactions							
Giving affectionate gestures	Husband	.95	19	.705	1.08	19	.692
	Wife	1.12	25	.600	1.14	21	.615
Giving violent gestures	Husband	.25	12	.622	.12	12	.311
	Wife	.50	16	.816	.41	17	.593
Receiving violent gestures	Husband	.55	11	.688	.31	13	.751
	Wife	.29	14	.469	.22	16	.515
Total	Husband	.76	318	.613	.80	333	.691
	Wife	.77	392	.679	.63	382	.624

Note: *The mean is between 0 (not enough) and 2 (a lot of the resource concerned), with 1 representing enough of the resource concerned, i.e., enough according to a personal benchmark of optimality.

functioning over time. Interestingly, when looking at the total means as perceived by spouses, wives tend to slightly overestimate husbands' optimality levels, with a mean of 0.80 against 0.76 from husbands' point of view, while husbands tend to underestimate wives' optimality levels, with a mean of 0.63 against 0.77 from wives' point of view. Similarly, "gendered" items describing housework, such as cooking/cleaning and repairs/driving exhibit gender biases. Husbands recognize that they are doing less cooking/cleaning than they should ideally do, and are doing nearly enough repairs/driving, while the opposite is true for wives, who recognize that they are doing nearly enough cooking/cleaning, and less repairs/driving than they should ideally do. The gender identity of the ideal household driver here is male, and the ideal household cook is female. It is plausible that, in this case, though the perception of optimality in housework is gender parity, this optimality is constrained by social norms dictating gender roles. These norms reinforce the interdependence of opportunity sets whereby gender norms are assigned to a specific allocation of resources. At the household level, social norms strongly influence the household optimal resource allocation. Optimality in resource allocation can be related to gender and cultural identities. However, social identities then have repercussions regarding the individual access to the labor market where the social perception of identities is reproduced. As described above, the aspiration of both husbands and wives for wives to access the labor market is not matched by an objective reality where gender inequalities still persist in the labor market. Ideals attached to gender and cultural norms are predominant in influencing both subjective well-being and access to individual and collective functionings, since they set the benchmarks for behavioral norms within and outside the household.

5.5. Conclusion

In terms of the social entitlements of *maquiladora* workers, as Mexicans, they inherit a rich cultural tradition that influences behavioral norms in all institutions. From governing bodies to the labor market and the household, norms and identity-related ideals are at the heart of the social organization for the allocation of resources and the distribution of income. In the past two decades, this organization has been challenged by changes in economic entitlements for the Mexican population due to their economic integration with North America. Traditional Mexican values have been affected, particularly the traditional household structure. Single-headed households and divorce rates have risen, and marriages and fertility rates have fallen. Female labor participation has increased in line with GDP growth, and the entire economy has begun to show signs of a newly developed country rather than a developing country. However, the health indicators suggest that economic integration has overtaken human abilities to absorb such changes. Mortality rates caused by diabetes, liver diseases, and malignant tumors have increased at a dramatic pace in the past few years. New patterns of consumption have begun to affect the functionings of a part of the population, especially the growing middle class. Economic integration also implies that a nation is not an independent economic entity but is integrated into and dependent on a global interdependent system. In 2001, the slowdown of the US economy resulted in a slow down in the economic opportunities for the Mexican economy as well. Finally, economic integration is also linked to a fusion of identities that spreads beyond borders, and this has created an American culture in Mexico, with the flows of FDI and commodities, and migration of Mexicans into the United States. In the United States, the broadly defined Hispanic identity has been spreading across the US labor market, legally or illegally.[11] In Mexico, the inflows of foreign investments led to better access to goods and services and increased living standards. However, in the midst of this process of integration, the fusion of identities has become a clash of identities in the Mexican border region, Ciudad Juarez being the mirror of this economic integration without a human face.

Chapter 6

Economic Entitlements

In the last two decades of the twentieth century, the economic performance of Mexico was essentially achieved through the expansion of trade and export-oriented industries after import-substitution industrialization from the 1960s, which saw the birth of the *maquiladora* industry. From 1983 to 1993, after the peso devaluation of February 1982, the annual growth rate of GDP was around 3 percent, while GNI per capita in current US$ was $2,270 in 1983 and rose to $4,230 in 1993, attaining $8,340 in 2007.[1] The public debt, including both internal and external debts, represented 62 percent of GDP in December 1983 against 20 percent of GDP in December 1993.[2] By the late 1980s, the government had adopted a strategy of economic liberalization and export-led economic growth. Coming into effect on January 1, 1994, the North American Free Trade Agreement (NAFTA) between the United States, Mexico, and Canada represents a further step in the process of North American economic integration. It was hoped by its advocates that NAFTA would bring new economic opportunities for Mexico in terms of employment and balance of trade by increasing the level of Mexican exports. On that same day, the Ejército Zapatista de Liberación Nacional (EZLN) began its insurrection in Chiapas to denounce in part the unequal redistribution of the fruits of growth (Dussel 2001). In December 1994, a devaluation of the peso hit the Mexican economy. Because it raised inflation, this devaluation put further pressure on the purchasing power of poor Mexican households. It was argued that the post-1995 economic growth of Mexico was to have benefited a growing middle class, multinational companies, and large national companies—including Telmex, for telecommunications, and the state-owned Pelmex, for oil production, which would benefit at the expense of small and medium enterprises (Dussel 2001).

The *maquiladora* industry is an important feature of the economic growth of Mexico. It provides employment on a large scale, stimulates industrial development, and attracts foreign investment while increasing foreign exchange earnings. In terms of economic entitlements, the *maquiladora* worker has access to a job position together with a real wage, which is a determinant of the worker's purchasing power. Mollick and Wvalle-Vazquez (2006) have shown that over the period 1990–2001, *maquiladora* employment did not fall as Mexican wages increased relative to Chinese wages. However, China's entry into the WTO in November 2001 was expected to put further pressure on wages and the output prices in the *maquiladora* industry. As Sargent and Matthews (2004) have argued, if *maquiladora* firms were to remain competitive with Chinese labor-intensive products, one strategic choice was to switch to more capital-intensive products. Another strategic choice, shown in Charles (2011a), is to increase the gender gap, lowering the female workers' wages relative to that of males, rather than reducing employment, in order to keep productivity constant or positive while cutting labor costs. The rationale behind this is that the gender wage gap in the *maquiladora* industry externalizes social norms of the paternalistic household. From the firm's perspective, a relatively lower wage for women's work is not discriminatory, given that this lower income entitlement for women is socially acceptable. This chapter therefore draws from these results and expands the analysis to show how consideration of identity affects the determination of wages paid to line-workers, male or female, and the productivity of the group as a whole. Using an identity model (Akerlof and Kranton 2010), this chapter argues that the behavior of principal and agent in the *maquiladora* industry is influenced by the identity to which they belong, and especially to the related social norms and ideals attached to their identity. Consequently, changes in wages, productivity, and social interactions at the workplace are influenced by identity-related norms of behavior.

6.1. Insider-Outsider Model with *Maquiladora* Identities

The level and quality of effort one is able to provide for a given wage depends on personal endowments and functionings. In effect, the effort provided by a worker depends not only on his or her educational skills, physical ability, and mental ability, but also on personal experience related to previous choices over limited opportunities. This is called the effort-related capabilities. The social perception of the effort-related capabilities provided by a group of workers sharing a personal identity such as gender, race, or age can serve as a basis for wage discrimination. Norms bias the decision-makers' perception of the effort-related capabilities of workers, and therefore also the level of wages paid to these workers. Thus, the asymmetric information regarding effort between employers,[3] employees,[4] or employers and employees[5] may lead to a normative judgment that serves as the basis of wage offers. As discussed in chapter 2, the problem of asymmetric information may in effect lead labor market actors to make decisions according to a hierarchy of ideals and associated behavioral norms. The model presented now explains the problem of asymmetric information as a matter of identity's stand point, whereby employers and employees make decisions according to the identities to which they belong, and especially according to the related norms and ideals attached to their multiple identities. The model draws from the identity models in Akerlof and Kranton (2010), thus following similar steps of analysis.

The Procedure: Part I. In the model, the *principal* is the employer, and the *agent* is a worker. The agent exerts effort according to his or her abilities and work incentives. Productivity depends on the quality and quantity of effort, that is, the ability and intensity required to perform certain tasks that engender more revenue or less revenue for the firm. The principal cannot observe the agent's effort, but he or she can observe whether revenues are high or low. The principal can influence the agent's effort by paying a higher wage if revenues are high than if revenues are low.

The Procedure: Part II. How will consideration of identity affect the level of wage paid to the worker, whether woman or man, and the productivity of the group as a whole? The wage paid by the principal will be derived from the perceived effort-related capabilities of the agent, whether insider or outsider, dictated by social norms, which will in turn affect productivity. Similarly to Akerlof and Kranton (2010, 42), the three identity ingredients are added to the model, namely, social categories, norms and ideals, and identity utility. However, the novelty here is to augment social categories with multiple identities, in line with the E-mapping theory.

Social Categories and multiple identities. The agents can be classified into two types. Those who identify with their firms or organization are called *insiders*. Those who do not identify with the organization are called *outsiders*. Insiders and outsiders are therefore the social categories of the model. A further degree of complexity in the present model is about considering that insiders and outsiders can display a similar social identity, whether male or female.

Established in the 1960s, the *maquiladora* industry was traditionally hiring female workers, because of their "nimble fingers" (Elson and Pearson 1981), for the assembly of manufacturing products. Consequently, female workers have been considered the insiders of the present model. However, since the peso devaluation in December 1994, the demand for *maquiladora* workers has increased at a fast pace, which meant that male workers, the outsiders of the present model, were increasingly hired to cope with the sudden rise in labor demand. Figure 6.1 shows a peak in workers' employment in 2001 after a dramatic increase in employment since the peso devaluation and NAFTA implementation in 1994, both of which triggered the demand for Mexican exports, including the demand for *maquiladora* products. Since the majority of those products are exported to the United States, it is likely that the 2001 recession in the United States was one of the main reasons for a sudden fall in employment level in the same year.

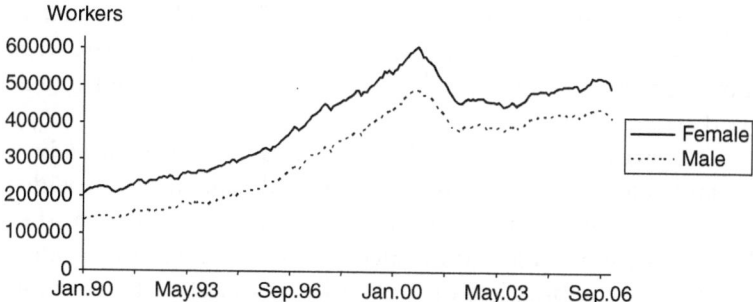

Figure 6.1 Employment trend of line workers at the national level by gender (1990–2006)
Source: INEGI (2011).

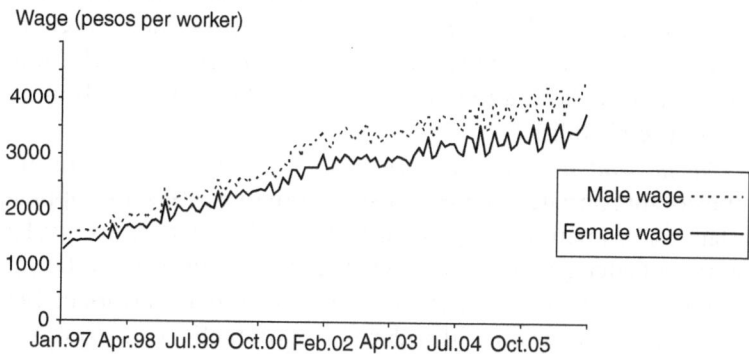

Figure 6.2 Nominal wages for *maquiladora* line workers by gender (1997–2006)
Source: INEGI (2011).

In the same job position of *maquiladora* line-workers, male and female workers are entitled to different wages. Based on monthly data from January 1997 to December 2006, figure 6.2 shows the nominal wages for *maquiladora* line workers by gender. Interestingly, the gender data is available from January 1997 to December 2006, since, as stated by INEGI (2011), the data cannot be updated for reasons related to the base year used, the conceptual methodology used, and simply because the information cannot be generated anymore. An important insight gained from the graph is about the size of the gender gap: around 10 percent in January 1997, which increased to around 15 percent in December 2006. The increase in the wage gap suggests a change in the relative cost of male and female workers.

The process of wage determination can be influenced by a social value attached to the effort-related capabilities provided by workers, depending on the social group to which they belong. In low-skilled jobs, effort is used to measure productivity, but the effort itself is not easily measurable. In assembly industries especially, teamwork constitutes an essential factor of productivity. Alchian and Demsetz (1972) argued that team production may make effort monitoring necessary, because an individual performance is not observable. Given that effort is not easily measurable, the monetary compensation for the abilities to work, for which workers are hired, is assumed to rely in part on a social value concerning which abilities are worth more than that of others, and which group's abilities are worth more than others' abilities.

Norms and ideals. The positions of principal and agent are labeled appropriate for men in some organizations and appropriate for women in other organizations. An agent is an insider or an outsider depending on whether the agent is a woman or a man in a particular organization. An insider who identifies with the organization will seek to exert high effort. In contrast, an outsider will think that the effort should be minimal—the outsider does not think about the productivity of the organization as a whole. As a work incentive, the outsider will have to receive a wage strictly superior to the insider's wage. Consistent with the internal reference perspective of efficiency wages,[6] the outsider will not exert the maximum effort-related capabilities for the organization unless the wage received is superior to the insider's wage. Through peer comparison, the insider and outsider make sense of the wage-effort combination required in their job position and across other job positions.

A social norm defining the effort-related capabilities of the insider as less productive and therefore less valuable than the outsiders' effort-related capabilities is discriminatory but may be perceived as socially acceptable. At the firm level, this is not discriminatory if lower income entitlement for women at the household level is socially acceptable. In effect, according to Mexican norms, women must have a lower reservation wage than men to avoid intra-household conflicts. A survey of Mexican values conducted in 2000 found that 61.1 percent of Mexicans agreed with the statement: "If a woman earns more money than her husband, it's almost certain to cause problems" (World Values Survey 2006). Such norms can be exacerbated in times of rising uncertainty such as a recession; men should still be placed in a better position than women in the labor market to be perceived as the main breadwinner. In 2001, China's entry into the WTO and the US recession put pressure on *maquiladora* workers' wages. The result was an increase in the gender wage gap, as described by figure 6.1. Charles (2011a) used Akerlof and Yellen's (1990) fair wage–effort hypothesis to explain the gender wage gap as a matter of "fair-wage constraints" that differ across genders, and which are due to evolving social norms of fairness in reservation wages for men and women within households. Empirical evidence for changes in gender wages gaps across industries between 1997 and 2006 is found to be consistent with this argument. Consequently, in the context of increasing demand for *maquiladora* workers, the perceived returns on female productive characteristics evolved over time, leading to a decrease in the cost of female labor around 2001–2002. From the employers' point of view, productivity is kept positive thanks to the positive growth rate of the wages of both men and women, which triggers higher efforts from both male and female workers. This also allows firms to keep hiring male workers and to pay them below the maximum wage that can be set in order to remain output-price competitive. The norm attached to women's effort-related capabilities has a lower value than male effort-related capabilities, based on the Mexican gender roles. This means that male workers, the outsiders, are entitled to a higher wage rate than female workers, the insiders, whose wages are adjusted according to the level of demand for the industry output.

Assuming that effort is not measurable and varies across individuals, the assessment of the cost of a particular type of workers must rely to some extent on the social perception of this type of workers outside the factory plant. Empirical evidence from the *maquiladora* industry in Mexico supports the argument that the concept of fairness, based on social norms, influences both employees' and employers' perception of the fair-wage level for groups of workers sharing a social identity: which here is gender. In the case of the *maquiladora* industry, social norms place a lower value on female effort relative to male effort. The gender wage gap in this industry is sustained by the role played by individuals within the household as mothers, fathers, daughters, sons, and in-laws. These roles are then perceived as social identities. The larger the gender wage gap within the household, the larger the gender wage gap in the labor market: the household income gap, to a certain extent, reflects the corresponding wage gap in the labor market. In other words, the norms within the household mean that there is a certain level of relative income between members. This level of relative income is then sustained in the labor market, where decision-makers ensure that every worker gets the relative wage dictated by the social norms of relative income.

Identity Utility. Based on observations, Akerlof and Kranton (2010, 87) argued that "women lose utility from working in a man's job. And men lose utility from working in a woman's job. Men also lose utility when a woman works in a man's job. They can also sabotage the work of women; this sabotage increases the perpetrator's utility but leads to lower productivity for everyone." However, if jobs are socially attributed to be men's jobs or women's jobs, the issue would simply be that a social identity loses identity utility whenever threatened by another identity in socially assigned occupations. Translated to the present insider-outsider model, it means that an insider loses identity utility when putting in lower effort than the ideal set by social norms, that is, a lower effort than the perceived effort-related capabilities appropriate for the job position. An outsider loses identity utility from working in an insider's job. For an outsider, to work in an insider's job is to be perceived as having similar effort-related capabilities, which creates a loss of identity utility. Sabotage occurs when there is a loss of identity utility, that is, when the insider's identity or the outsider's identity is threatened. Sabotage will occur until social norms evolve to an ideal where the position of agent is labeled appropriate for both men and women. Sabotage of the agent's working abilities can take different forms, from effort withdrawal and bullying behavior, to sexual harassment and violent behavior.[7]

Ideals. Each identity is related to an ideal that influences the behavior in each social category. Insiders and outsiders target an ideal level of productivity according to effort-related capabilities appropriate for the job position. Men and women target an ideal level of wage according to their effort-related capabilities. In society where women are socially entitled to a lower income than men, a male outsider will have to receive a strictly higher wage than the female insider's wage, as a work incentive.

6.2. Trends in *Maquiladora* Labor Market

Between 1990 and 2006, the *maquiladora* industry was affected by its close relationship with the US economy and by the accession of China to the WTO in November 2001. From 1995 to 2000, the increase in US demand for *maquiladora* output led to an increase in the labor demand for *maquiladora* workers. From 2001, downturns in the US economy reduced output in Mexican *maquiladora* sectors, and China's entry into the WTO increased competition for the *maquiladoras'* output of labor-intensive products. Sargent and Matthews (2004) note that one consequence of increased competition was a change of industry structure, from mainly labor-intensive outputs to progressively more capital-intensive outputs. Furthermore, the *maquiladora* industry has seen a rise in the proportion of male workers and an increase in the gender wage gap over the past two decades. The extent to which these phenomena are linked needs to be explored looking at the evolution of male employment by each sector of the industry.

Male Share of Employment

Both demand-side and supply-side factors are responsible for gender differences in economic outcomes.[8] Catanzarite and Strober's study of *maquiladora* workers in Ciudad Juarez asserts that a number of demand-side factors have affected the growth in men's and married women's share in *maquiladora* employment; these factors include the deterioration of the Mexican economy after the 1994 currency devaluation, relatively improved wages in other sectors, and an emphasis on the complexity of tasks (Catanzarite and Strober 1993). Two separate studies also argue that the "defeminization" of the *maquiladora* labor force since the implementation of NAFTA in 1994 has come about as a result of labor demand for workers in the *maquiladora* industry growing faster than the absorption of the working-age population into the labor force (Fussell 2000; Fleck 2001). Consequently, the "reserve army" of people who were not previously employed in the *maquiladoras*, such as men and married women, filled the spots that had been assigned to young single women in the past. Changes in the *maquiladora* program and the expansion of output and employment in more capital-intensive industries, such as transportation equipment, have led to shifts in employment. For example, when *maquiladora* laws expanded production, relatively more men worked in *maquiladoras*, given the lack of job opportunities for unskilled workers outside the informal sector. Over a couple of decades, job opportunities expanded in the heavily capitalized transportation equipment sector, from 6.2 percent of all *maquiladora* workers in 1980 to 18.7 percent in 1998 (Fleck 2001). In the post-2000 era, heavy competition attributed to China's entry into the WTO in 2001 created acute competition for labor-intensive products. This factor, as well as the US recession of 2001, led to a reduction in the demand for *maquiladora* output.

From 1990 to 2005, changes in the ratio of female to male workers in the *maquiladora* labor force should not be considered as a fall in female employment, which actually increased by 146 percent over this period (table 6.1). Looking at table 6.2, since 2001, the proportion of total employment accounted for by the traditional *maquiladora* sectors, such as textiles, transport tools, and electronic products and accessories, has fallen by 5 or 6 points. However, in 2005, these sectors still remained the main sectors of *maquiladora* employment. The textile sector has seen its employment level rising by 353 percent over the period from 1990 to 2005 (table 6.2); the proportion of male workers increased by

Table 6.1 Employment proportion and growth of workers per output type and gender* in the *maquiladora* industry (percentages)

Sector per type of output	1990 M	1990 F	1995 M	1995 F	2001 M	2001 F	2003 M	2003 F	2005 M	2005 F	1990–2005 M	1990–2005 F
Food processing/packaging	39	61	44	56	49	51	50	50	52	48	+61	−3
Textiles	24	76	29	71	39	61	42	58	43	57	+702	+240
Leather products	47	53	46	54	48	52	53	47	57	43	−9	−38
Wood/metal products	71	29	70	30	69	31	68	32	66	34	+130	+183
Chemical products	47	53	46	54	47	53	45	55	45	55	+320	+362
Transportation equipment	47	53	50	50	52	48	51	49	50	50	+175	+149
Tools assembly/repairs	64	36	60	40	63	37	57	43	56	44	+280	+428
Electric/electronic products	37	63	37	63	44	56	46	54	46	54	+192	+101
Electric/electronic accessories	32	68	35	65	39	61	39	61	39	61	+184	+112
Toys/sport products	33	67	39	61	47	53	46	54	45	55	+1	−38
Other types of products	40	60	44	56	46	54	46	54	46	54	+312	+234
Services	25	75	31	69	41	59	45	55	46	54	+273	+44
Total absolute number of workers	1553309	2419443	2578716	3735098	5277040	6414882	4636848	5460582	5079694	5952396	+160264	+103083
										Total	+ 227	+ 146

Source: Table reproduced from Charles (2011a).

Note: *M=male workers' sum, and F=female workers' sum; up to 100 for each sector each year.

Table 6.2 Employment proportion and growth of workers per output type in the *maquiladora* industry (percentages)

Sector per type of output	1990	1995	2001	2003	2005	1990–2005
Food Processing and Packaging	2	1	1	1	1	+22
Textiles	11	16	23	21	18	+353
Leather products	2	1	1	1	0.1	−24
Wood/metal products	5	6	5	5	5	+145
Chemical Products	2	2	2	2	2	+342
Transportation equipment	23	21	28	22	22	+161
Tools assembly and repairs	1	1	1	2	2	+333
Electric/electronic products	11	1	7	9	1	+134
Electric/electronic accessories	25	25	24	22	21	+135
Toys and sport products	2	1	1	1	1	−25
Other types of products	11	11	12	13	15	+266
Services	6	4	4	3	4	+102
Total maquiladora industry	100	100	100	100	100	+178

Source: Table reproduced from Charles (2011a).

702 percent, while the proportion of female workers increased by 240 percent over this period (table 6.1). Technological change means that computer-aided marking and grading, semiautomatic sewing and pressing machines, and computer-controlled cutters have increased the productivity per worker, while it also increased the need for a qualified labor force and reduced the need for low-skilled labor (Infomat 2000; 2006). The same trend in the ratio of female to male workers is also evident in the transport tools and accessories sectors, where male workers increased by 175 percent against a 149 percent increase for the female labor force over the period from 1990 to 2005. Again, the same trend occurs in one of the three lynchpins of the *maquiladora* industry: the sector concerning electric and electronic accessories. Finally, the sector defined as "other types of products" has seen its share of the total *maquiladora* industry increase by 266 percent from 1990 to 2005. Given the increase in the number of both male and female workers in this sector—312 and 234 percent, respectively, over the same period—it can only be assumed that the types of products concerned follow the same trend of more capital-intensive products. However, the largest increases in female employment were in sectors requiring skilled labor. In the sector related to tools assembly and repairs, female employment increased by 428 percent from 1990 to 2005, with female share increasing from 36 percent of the labor force in 1990 to 44 percent in 2005. Similarly, in the sector related to chemical products, female employment increased by 362 percent from 1990 to 2005, with female share increasing from 53 percent of the labor force in that sector in 1990 to 55 percent in 2005. Therefore, the evidence is not clear to support the argument that male employment increased at the expense of female employment due to an increase in demand for a higher-skilled labor force. The evidence is rather that female workers had to go to niche sectors representing 3 percent of total employment in the *maquiladora* industry.

Although real wages slipped significantly in the 1980s and in 1994 due to inflationary pressures resulting from currency devaluations in 1982 and 1994, nominal wages in the industry have risen steadily over the past 15 years (INEGI 2011). For instance, in 1994 the real minimum urban wage was only 42 percent of its 1980 value (Chant 2003). Following the December 1994 devaluation, a similar pattern arose. In labor supply terms, this reinforces the view that more people (particularly mature women) have to look for jobs to sustain household living standards in the context of inflationary cost of living, and this is particularly relevant in the *maquiladora* industry, which offers a stable low wage and a range of benefits.[9] Since everyone experiences an absolute decline in the value of wages, there should be an increase in labor supply across the board. Socio-demographic factors in effect show significant signs of increasing labor supply. As shown in the previous chapter, over the period from 1990 to 2005, the divorce rate in Mexico increased by 65 percent (from 6.6 percent of all marriages in 1990 to 11.2 percent in 2005), marriages declined by 7 percent and fertility rates dropped by 35 percent, from 3.4 children per woman in 1990 to 2.2 children per woman in 2005. These figures suggest that more female labor, previously occupied in the household economy mainly, may have joined the overall labor force. In effect, female participation in the labor force increased from 36.9 percent of the working-age population in 2000 to 41.5 percent in 2007. This increase in

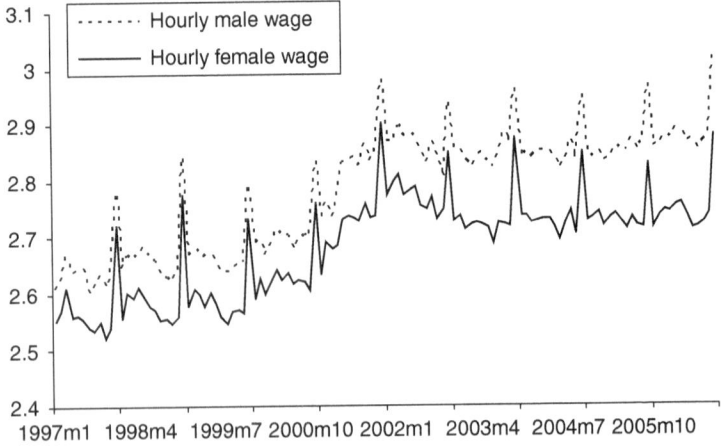

Figure 6.3 Log of the gender wage gap in the *maquiladora* industry

Source: Figure reproduced from Charles (2011a). Hourly male or female real wage is calculated as the real wage per male or female worker divided by the number of working hours per male or female worker on average.

the overall labor supply was most significant in the labor market for low-skilled workers. For example, female participation in agriculture and fishing sector increased by 169 percent over the period from 1998 to 2003, while the corresponding increase for males was 15.5 percent. Changes in norms influence the level of relative wages and the perceived level of effort associated with the identity of the workers hired. In that context, changes in the notion of fairness within Mexican gender norms are investigated via the analysis of economic pressures on the gender wage gap and the labor demand for *maquiladora* workers.

Trends in the Gender Wage Gap

Based on monthly data available at the industry level only, figure 6.3 depicts the gender wage gap between the hourly wages of female and male workers in the *maquiladora* industry from January 1997 to December 2006. Apart from the presence of seasonality, an important insight from the graph is the sudden change in trend around 2001, which supports, albeit only graphically, the role of Chinese competition and/or the US recession in affecting the gender wage gap. As mentioned above, the size of the gender gap was around 10 percent in January 1997 and increased to around 15 percent in December 2006. The emphasis here is as much on the sudden change in trend as it is on the presence of a wage gap and the socially perceived relative value of workers' effort. The reason put forward for such trends and breaks in trends is that social norms within the profit-maximizing firm and within Mexican society need to translate into two fair-wage constraints in the labor market for male and female workers.

On the labor supply side, once female members decide to enter the labor market, it is socially acceptable for them to bring an income into the household as long as it is not as high as the income of their male counterparts. In other words, the reservation wage for women must be lower than for men at the household level, which translates into two fair-wage constraints in the labor market. Wage discrimination theories often rely on the statistical fact that women display higher labor supply elasticity than men, rather than on perceived differences in productivity between men and women. Increasing product market competition reduces gender discrimination, while a larger share of women workers is associated with stronger performance in the United States and the UK (Belfield and Heywood 2006). In Mexico, however, the opposite seems to be the case: increasing product market competition through global competition or intra-industry competition raises gender wage discrimination in the home country. This is due to the norms attached to the reservation wage of women's identity within the household and the resulting fair wage constraint, which must be higher for male workers than for female workers in the labor market. Although it reflects universal practices of identity wage discrimination, these norms are specific to this context of the Mexican gender *maquiladora* identity, where the cultural background reflects strong traditional religious and ancient indigenous beliefs. Related norms and values have facilitated the sustainability of a patriarchal system. Therefore, behavioral norms derived from a long history of cultural practices seem to underpin a patriarchal system that, it could be argued, partially explains the workers' access to wages and employment.

Money wage or current disposable income is a major issue when considering individuals' perceptions of their position within the household or in the labor market. In 2005, the inactivity rate (in terms of participation in the paid labor market) for Mexican women aged between 25 and 64 years was around 50 percent, while the rate for Mexican men of the same age group was around 3.5 percent (KILM 2005). The corresponding female inactivity rates in the United States and the UK are around 10 and 22 percent, respectively. Male breadwinners must remain the main source of money earnings and must always earn more than their female counterparts. This norm is reinforced in times of economic hardship, since additional household labor resources are pulled into the market in order to sustain the same living standards. Thus, workers will not ask for more than the share to which they are socially entitled in comparison with the other gender. Lower reservation wages for women are socially accepted and therefore constitute the standard of economic entitlement for female workers entering the *maquiladora* labor market. The weight of the informal sector, representing around 30 percent of total employment for both men and women over the past 20 years, and high labor turnover in that industry help to sustain the low bargaining power of *maquiladora* workers.

6.3. Social Norms, Wages, and Productivity

Finally, the last step of the analysis to show how norms can affect both wage determination and productivity is to look at the correlation over time between changes in the wages and productivity variables of the *maquiladora* industry. In other words, by looking at the covariance between changes in productivity, changes in the wages of male workers, and changes in the wages of female workers, the interdependence of wage changes between male and female workers should highlight the norm of fairness concerning which identity is entitled to increasing wages linked to increases in productivity. The monthly time series data set is retrieved from INEGI (2011) to look at the long-run relationships and covariance between the real wage of female workers (*frw*), the real wage of male workers (*mrw*), and the level of productivity of the entire industry (*pty*) over the period from 1997 to 2006. For reasons of data availability mentioned before, the size of the data set is of 108 observations from January 1997 to December 2006 only. The variable *pty* representing productivity has been calculated by the author as the value added in real terms (Mexican pesos) over the number of hours worked per month for all *maquiladora* workers, male and female. Value added is that generated for exports, which includes all the costs of production, such as social benefits, employers' contribution, overall salaries, raw material, packaging, capital expenses, and gross profits (INEGI 2011). The hourly real wage for male workers *mrw* has been calculated as the real wage per male worker in pesos[10] divided by the average number of hours worked per male worker, and similarly for the calculus of the hourly real wage per female worker *frw*. The series are logged and seasonally adjusted given that signs of productivity peaks are detected in December of each year. Figure 6.4 shows the three series used in the analysis of covariance after seasonal adjustment.

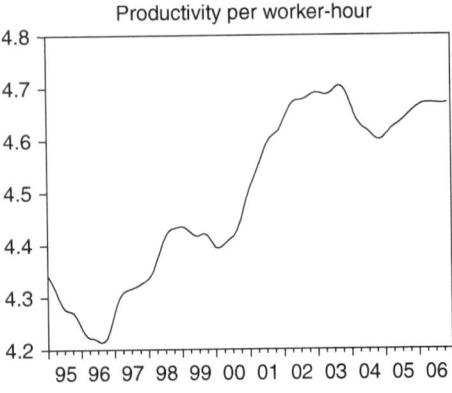

Productivity per worker-hour

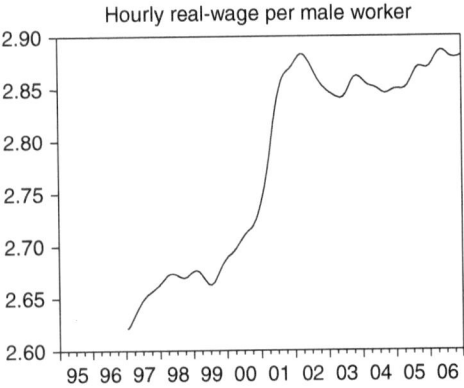

Hourly real-wage per male worker

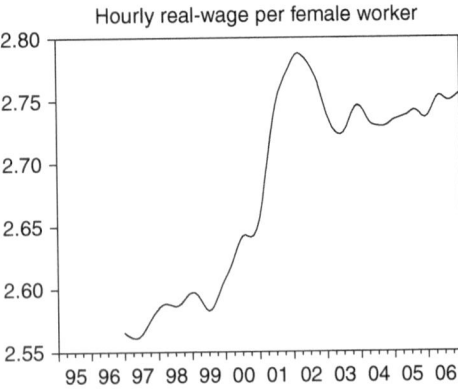

Hourly real-wage per female worker

Figure 6.4 Log of productivity and real wages over ten years (1996–2006)

Table 6.3 Variance decomposition of productivity, real wage per male and female worker

Variance decomposition of productivity (*pty*)

Period	S.E.	*pty*	*mrw*	*frw*
1	0.000	100.00	0.00	0.00
2	0.003	99.89	0.06	0.04
3	0.006	99.69	0.18	0.13
4	0.01	99.45	0.33	0.21
5	0.01	99.26	0.47	0.27
6	0.02	99.13	0.59	0.28
7	0.03	99.06	0.69	0.25
8	0.03	99.01	0.79	0.21
9	0.03	98.9	0.92	0.17
10	0.03	98.71	1.15	0.14

Variance decomposition of real wage per male worker (*mrw*)

Period	S.E.	*pty*	*mrw*	*frw*
1	0.001	11.83	88.17	0.00
2	0.002	15.10	84.87	0.02
3	0.004	18.93	80.95	0.11
4	0.006	23.34	76.36	0.29
5	0.009	28.25	71.21	0.53
6	0.01	33.53	65.72	0.75
7	0.01	38.97	60.18	0.85
8	0.02	44.35	54.85	0.79
9	0.02	49.44	49.91	0.65
10	0.02	54.05	45.43	0.51

Variance decomposition of real wage per female worker (*frw*)

Period	S.E.	*pty*	*mrw*	*frw*
1	0.000	4.75	34.11	61.14
2	0.001	6.22	39.78	53.99
3	0.003	8.42	43.31	48.27
4	0.006	11.48	45.16	43.36
5	0.008	15.46	45.70	38.84
6	0.01	20.3	45.27	34.42
7	0.01	25.81	44.15	30.04
8	0.02	31.62	42.61	25.77
9	0.02	37.33	40.82	21.85
10	0.02	42.61	38.93	18.46

Table 6.3 then shows the variance decomposition of each variable according to the three variables under study. The variance decomposition of productivity shows that its variance results to a large extent from previous levels of productivity. Concerning a possible trade-off effect between the cost of male labor and the cost of female labor, productivity seems to be increasingly affected by the variance in the cost of male workers in the previous year, up to the tenth month, but in a marginal fashion. This is confirmed by the second variance decomposition of the real wage for male workers, which shows that previous levels of both productivity and male workers' wage affect the current level of male workers' wages. The real wage of female workers has very little implication in the variance decomposition of male workers' wage. However, looking at the third variance decomposition of the real wage for female workers, the picture is strikingly different. In effect, the variance in decomposition of the female labor cost is increasingly related to both productivity and the male cost of labor in previous periods, while previous levels of wages for female workers is less and less related to its current level. From table 6.3, it can be seen that the cost of female labor depends on the pay rate for male workers, which itself depends on previous levels of productivity, especially in previous periods 8, 9, and 10.

Following these results, a simple causality test between the variables under study was performed. The results are displayed in table 6.4 to show the extent to which changes in one time series can explain the variance of the other two series. The table shows that the results for the changes in productivity are not

Table 6.4   Granger causality/Exogeneity Wald tests

	Dependent variable		
	Δpty	Δmrw	Δfrw
Δpty	–	8.19 [0.04]	4.99 [0.17]
Δmrw	3.38 [0.34]	–	26.33 [0.00]
Δfrw	1.44 [0.69]	8.52 [0.03]	–
All	3.46 [0.75]	14.07 [0.03]	31.71 [0.00]

Note: The coefficients are Chi-square values for the tests, with probabilities in brackets.

statistically significant when looking at the probabilities values given in brackets. Other factors should be included to make significant inferences, such as capital or a time trend. The focus of this analysis is, however, to highlight a social norm on effort-related capabilities by looking at the changes in the cost of female and male labor. The main significant results are, therefore, that, on one hand, a change in the real wage for male workers is equally derived from changes in productivity, with a coefficient of 8.19, and from changes in the cost of female labor, with a coefficient of 8.52, both significant at the 5 percent level. On the other hand, changes in the real wage for female workers depend to a large extent on changes in the real wage for male workers, with a coefficient of 26.33, which is significant at the 5 percent level. Changes in productivity are not statistically significant in causing changes in the cost of female workers.

6.4. Conclusion

The post-NAFTA economic integration strengthened Mexico's link with the US economy and led to increased work and wage opportunities for current and potential *maquiladora* workers from 1995 to 2000. However, in 2001, the increase in competition from the low-paid labor force in China and the slowdown of the US economic activity led to a sudden rise in the level of the gender wage gap between *maquiladora* line workers. The significance of China's entry into the WTO in determining the level of male or female employment is difficult to assess due to its chronological correlation with the US recession, but this factor certainly contributed to the *maquiladora* industry becoming more capital-intensive. The increasing gender wage gap comes from the lower reservation wage for female unskilled workers than for male unskilled workers, which is then conveyed by social norms in the labor market by two fair-wage constraints. An evolving norm of fairness according to the economic environment means that the social perception of reservation wages changes in the event of economic downturns. Female workers have a lower reservation wage than male workers, which can then be adjusted according to the economic pressures on the reservation wage of male workers. The alternative for unskilled workers is no wages in the unpaid economy or an insecure income from the informal sector. Consequently, in the context of increasing demand for *maquiladora* workers, the perceived returns on female productive characteristics evolved over time, which led to a decrease in the cost of female labor around 2001–2002. From the employers' point of view, productivity is kept positive thanks to the positive growth rate of both wages, which triggers higher effort from both male and female workers. This also allows firms to keep hiring male workers and to remain output-price competitive.

Conclusion

When James Galbraith (2011) raises the issue of economic inequality between different political regimes, he puts forward the explanation that economic inequality in the new global economy is linked to intersectoral differentials between regional blocks of production, rather than international differentials between national political powers. The global political economy is now based on the standard set by the profit maximizer in financial speculations, which has taken over national interests in social interactions. The great recession that started in the late 2000s reflects in many ways greedy behaviors, but not uniquely in financial markets. Accumulation can take many forms. As profit maximizers, the accumulation of financial assets led to financial turmoil. As utility maximizers, homeownership became a possibility with subprime loans, and better access to cheaper foods led to rising obesity across the globe. The global economic integration led to the convergence of consumption patterns due to peer comparison and habit formation, together with the creation of regional blocks of production. Economic integration goes through the production and accumulation of resources. However, the current crisis, which may be perceived as a crisis of the system of accumulation, is essentially a crisis of allocation of resources. In effect, limiting the human complexity to two identities, that of profit maximizers or utility maximizers, undermines the hierarchy of multiple identities valued in social interactions as a major determinant of inequality. The discussion in this book shows how standards set by social interactions influence to a large extent market interactions and the resulting allocation of resources between gender, race, and national and regional identities. Starting from the identity's point of view, E-mapping shows the effect of economic events on individual well-being and inequality. Social norms and associated ideals articulate the impact of such events on different groups' identities. The group being at the bottom of the hierarchy of ideals in social interactions is the most vulnerable in market interactions, in terms of both living standards and the development of capabilities.

Economic events affect the capabilities of individuals in a positive or negative way depending on their identities, not uniquely as consumers and wage earners, but also as household members or simply as human beings with a metabolism shaped by a specific geographical and historical environment. The geographical link between the United States and Mexico, tied to NAFTA, led to a North American product market that rapidly changed the cultural and social habits of Mexicans. Limited to the free trade of merchandises only, this economic integration strongly affected their human functionings. Over the past two decades there has been a significant increase in diabetes-related mortality linked to changes in consumption patterns, especially for the growing middle class. The access to the flow of goods and services depends on purchasing power. However, if this flow of goods and services is not followed by rising purchasing power for all consumers seeking social status, or for all breadwinners (male and female) seeking to preserve living standards, increasing violence and corruption (including drug wars), and also increasing migration across the border, may arise as a consequence. Lack of jobs and income opportunities and rising gaps in purchasing power between groups of consumers are also consequences of economic integration.

The interdependence between identities is a major determinant of individual well-being and economic outcomes. Individuals assess and reassess their state of well-being relative to each other, within and across identity groups, and then they tend to get used to new standards imposed by policy outcomes over time. The effect of economic policies on capability-related well-being depends on social norms, that is, the social norms setting the rules of access to commodities and potential functionings. In the case study presented in the last part of the book, female workers were negatively affected in terms of real wages and employment opportunities in comparison with male workers in the *maquiladora* industry. At the household level, by investigating household members' perceptions of each other, two major aspects of well-being are shown. First, the interdependence of household members in valuing each other's capabilities has a significant impact on the life satisfaction of individuals. Valuing one's functioning is, to a large extent, a function of others' valuation of one's functioning. The extent to which each individual answers to others' expectation of their socially assigned group identities is a major factor in determining subjective well-being. Individual's capability sets are interdependent according to group identities within and outside the household. Second, the presence of adaptive preferences or habit formation in individual behavior is the essence of human functionings, since it reflects the ability of adaptation to one's living environment. Economic and social entitlements determine access to potential functionings and commodities, and being adaptable to new entitlements is an essential human functioning in itself. It is from this adaptation to one's environment that new functionings emerge. The crucial point here, however, is the standard set by ideals according to which people are guided to adapt. For example, a sustainable environment is one standard that should be seated at the top of the hierarchy of individual ideals in order to ensure that behaviors in social interactions are according to this universal target of sustainability of the human species as a whole.

Subjectivity is an essential element to assess individual well-being, which goes beyond the evaluation of utility raised from the consumption of commodities, and which is not limited to the development of capabilities only. Individual functionings are influenced by the individual's multiple identities, which have more value or less value according to culture, social values, legal systems, or ethnic groups, for example. Social norms are therefore at the centre of entitlements and shape the identity E-mappings of an individual and his or her related economic entitlements, such as employment opportunities and access to income. Norms influence human capabilities and individuals' choices over market opportunities as much as they create entitlement failures to these opportunities according to a hierarchy of ideal behaviors in social interactions. By identifying the entitlement failures or lack of opportunities to reach potential functionings for one particular group, one can understand and implement policies to remove those failures. Entitlement failures for the group identity affected should ideally be identified and corrected by an internationally agreed institution whose ideal should be about assuring that human beings can function in their environment according to their multiple identities. In social interactions, nongovernmental organizations already work in that sense. In market interactions, however, a credible global political power embedding regional interests is more urgently needed to set a common ideal for human beings to exchange and function in a sustainable fashion.

Notes

Introduction

1. Steckel (2008).
2. Robeyns (2005b).
3. The notion of "entitlement" is used extensively by both Nussbaum (2003; 2006; 2011) and Sen (1981; 1999).
4. US Census Bureau (2006).
5. World Bank (2009).
6. Alarcon et al. (2007).

1 Subjectivity in Well-Being

1. Sen and Nussbaum are considered to be the legitimate founders of the CA, although the idea of capabilities emerged earlier in Sen's work (1985; 1992).
2. Van Staveren (2008) provides a thorough discussion of the pros and cons of Sen's and Nussbaum's views on the CA.
3. "Fuzzy" is understood in its mathematical sense, representing the fact that elements of a set can have different degrees of membership to the set ranging from 0 to 1, instead of the classical set theory where elements do belong to or do not belong to the set, taking the values 1 and 0, respectively.
4. See Kahneman (1986; 2000); Kahneman et al. (1986; 1999); Kahneman and Krueger (2006); Kahneman and Tversky (2000).
5. The idea of a set point was notably put forward by Kahneman and Tversky's (1979) prospect theory reference-point analysis.

2 Identity, Norms, and Ideals

1. In that sense, social norms are emotional and behavioral propensities of individuals (Elster 1989) to act according to their

identity-related optimal behavior, or in other words, how far each individual is from an identity-related ideal.

2. The literature looking at the role of social norms on market outcomes includes Fehr's substantial works (Fehr and Schmidt 1999; Fehr and Fischbacher 2002; Fehr et al. 1998; Fehr et al. 2009), Kahneman (1986), and Skott (2005).

3. Similarly to Burke and Peyton Young (2010), no distinction is made between social norms and social conventions. Part of this section draws on Charles (2011c).

4. Constructivist rationality, on the other hand, suggests that individual decisions are based on a conscious deductive process of human reason, which creates rules of action.

5. See also Davis (2011) for a thorough discussion of identity in economics. Part of this section draws on Charles (2011b).

6. Similarly to Davis' self-narrative capability mentioned above (2011), Sen (2005) argues that people also have a capacity to reflect on their individual set of identities.

7. In that respect, Woersdorfer (2010) argues that the act of consumption starts from signalling social status to a means of norm compliance. Looking at the norm of cleanliness in the nineteenth century, Woersdorfer in effect shows that consumer preferences are not entirely subjective, in the sense that consumers share certain motivations due to their genetic inheritance. Status-seeking consumption of previous generations for an ideal of personal hygiene has become the norm over time.

8. DeMartino (2010), for example, points out to the lack of ethics and the need for ethical behavior in the economics profession.

9. See, for example, gender disparities in access to capabilities in Beutelspacher et al. (2003), Qizilbash (1997), and Sen (1987; 1999).

3   Exchange Entitlement Mapping

1. The notion of "entitlement" is used extensively by both Nussbaum (2003; 2006; 2011) and Sen (1981; 1999).

2. See Carabelli and Cedrini (2009) for a discussion of Keynes on happiness and economics.

3. Basic capabilities in Sen's tradition refer to the real opportunity to avoid poverty, and are a subset of all capabilities (Robeyns 2005b, 101).

4. See Alkire (2005), Clark (2003), Clark and Qizilbash (2005), and Nussbaum (2006).
5. Assuming education and health are not privatized in that case.
6. See, for example, Djebbari (2005), Folbre (1986), Handa (1994), and Rangel (2006).
7. See, for example, Rangel (2006), and Duflo and Udry (2004).
8. See, for example, Handa (1994), and Sotomayor (2009).
9. See Kennedy and Peters (1992).
10. See Rader (1980), Gasper and Van Staveren (2003), and Iversen (2003).

4 External Shocks on E-Mapping

1. These authors include Rader (1980), Gasper and Van Staveren (2003), and Iversen (2003).
2. See, for example, Alarcon et al. (2007), Robertson (2005), and Seguino (2000).
3. See, for example, Dussel (2001), Ghiara (2001), Palma (2003), and Rodgers and Nataraj (1999).
4. See, for example, Gruben (2001) and Mollick and Wvalle-Vasquez (2006).
5. See, for example, Djebbari (2005) and Kennedy and Peters (1992).
6. See, for example, Easterlin (2005) and Rayo and Becker (2005) on adaptive and interdependent preferences.
7. Irrational preferences mean that the choices made need not to be consistent through time, which also defines bounded rationality (Conlisk 1996). Bounded rationality in that context would mean that someone is altruistic in some directions—to her or his own family, for example—but individualistic in other directions—to strangers could be another example—without any predefined order.
8. This is consistent with Neumann and Morgenstern's theory of risk aversion (1944), arguing that individuals experience diminishing marginal utility, that is, individuals dislike losing more than they like gaining.
9. Total household income also includes property income.
10. Banco of Mexico data accessed through INEGI website (INEGI 2011).

11. This does not undermine the fact that inflation and economic growth are positively correlated, but only for relatively low levels of inflation, from 15 to 18 per cent according to Pollin and Zhu (2006), which is far below Mexican inflation levels after 1994.

12. See Tymoigne and Wray (2006) for a modern approach to the history of money; and see Tcherneva (2006) for a Chartalist approach to money.

13. Benefits and working conditions are assumed to be part of the real wage and employment, respectively.

14. Then, at the local level, the firm adjusts the industry wage according to local unemployment and local competition. However, focusing on the industry level, the firm level is left aside in this research.

15. See, for example, Ehrenberg and Smith (2006), Borjas (2005), and Hamermesh (1993).

5 Social Entitlements

1. The data is taken from a sociological census by the National Institute for Statistics (INEGI 2011).

2. None of the humans sacrificed were eaten by other members of the Aztec communities, as humans were seen as sons of gods and therefore as bearing a sense of divinity. Eating human flesh would have been a sacrilege.

3. Chang (2007) in effect showed how Western economies fostered their own industrialization with protectionism and also showed that, through the standards set by the World Bank, the IMF, and the WTO, they were then expecting developing countries to foster industrialization with free-trade and laissez-faire types of policies.

4. These crimes are characterized by extreme violence and remain mostly unpunished (Amnesty International 2003).

5. See, for example, Robinson (1969), Carlin and Soskice (2005), Blanchflower and Oswald (1994).

6. See, for example, Zepeda (2008).

7. Mexican National Institute for Statistics (Instituto Nacional de Estadística Geografía e Informática 2011).

8. See, for example, Marjit et al. (2007), Rodgers (2004), Şirin Saracoğlu (2007).

9. Robinson's 1969 book was first written in 1933.

10. These levels are consistent with the measurement procedures of qualitative research methods, whose distinguishing features are identity, rank order, equal intervals, and true zero point (Herzog 1996; Kahneman and Tversky 2000).
11. For example, the earnings of Hispanic men overtook the earnings of black men in managerial and financial occupations since 2001, which signals the rising status of the Hispanic identity in high-earning occupations (Arestis et al. 2011).

6 Economic Entitlements

1. World Bank (2009).
2. Banco de Mexico (2008).
3. See, for example, Schönberg (2007).
4. See, for example, Akerlof and Yellen (1990), Fehr et al. (2009).
5. See, for example, Chang and Wang (1996).
6. Both Akerlof and Yellen (1990) and Danthine and Kurmann (2006) adopt the internal reference approach in determining the level of efficiency wage and the subsequent problems of wage rigidity in labor markets.
7. A similar analysis could be conducted with the position of principal, which would explain the Glass Ceiling effect.
8. See, for example, Blau et al. (2002).
9. For interesting sociological insights from the *maquiladora* industry, see Sklair (1989).
10. The price level used is the 2002 Mexican CPI (INEGI 2011).

Bibliography

Ahnert, L., Gunnar, M. R., Lamb, M. E., and Barthel, M. (2004). "Transition to Child Care: Associations with Infant-Mother Attachment, Infant Negative Emotion and Cortisol Elevations." *Child Development* 75: 639–650.

Akerlof, G. A. (1984). "Gift Exchange and Efficiency-Wage Theory: Four Views." *The American Economic Review* 74(2): 79–83.

Akerlof, G. A. and Kranton R. E. (2000) "Economics and Identity." *The Quarterly Journal of Economics* 115(3): 715–753.

Akerlof, G. A. and Kranton R. E. (2010). *Identity Economics: How Our Identities Shape Our Work, Wages, and Well-being*. Princeton: Princeton University Press.

Akerlof, G. A. and Yellen, J. L. (eds) (1986). *Efficiency Wage Models of the Labor Market*. Cambridge: Cambridge University Press.

Akerlof, G. A. and Yellen, J. L. (1990). "The Fair Wage-Effort Hypothesis and Unemployment." *Quarterly Journal of Economics* 105(2): 255–283.

Alarcon, D., Osorio, R., Soares Veras, F., and Zepeda, E. (2007). "Growth, Poverty and Employment in Brazil, Chile and Mexico." Working Paper (42), International Poverty Centre, UNDP.

Alchian, A. A. and Demsetz, H. (1972). "Production, Information Costs, and Economic Organization." *The American Economic Review* 62(5): 777–795.

Algan, Y., Dustmann, C., Glitz, A., and Manning, A. (2010). "The Economic Situation of First and Second-Generation Immigrants in France, Germany and the United Kingdom." *The Economic Journal* 120: F4–F30.

Alkire, S. (2002). *Valuing Freedoms: Sen's Capability Approach and Poverty Reduction*. Oxford: Oxford University Press.

Alkire, S. (2005). "Subjective Quantitative Studies of Human Agency." *Social Indicators Research* 74(1): 217–260.

Altman, M. and Lamontagne, L. (2004). "Gender, Human Capabilities and Culture within the Household Economy: Different Paths to Socio-Economic Well-Being?" *International Journal of Social Economics* 31(4): 325–364.

American Human Development Index (2010). *The Measure of America 2010–2011: Mapping Risks and Resilience.* Social Science Research Council.

Amnesty International (2003). "Mexico: Intolerable Killings: 10 Years of Abductions and Murders of Women in Ciudad Juarez and Chihuahua: Summary Report and Appeals Cases." Index Number AMR 41/027/2003.

Anand, P., Hunter, G., and Smith, R. (2005). "Capabilities and Well-Being: Evidence Based on the Sen-Nussbaum Approach to Welfare." *Social Indicators Research* 74(1): 9–55.

Aquinas, T. (2006). *Summa Theologiae: Man.* Cambridge: Cambridge University Press.

Arestis, P., Charles, A., and Fontana, G. (2011). "Financialisation, the 'Great Recession' and the Stratification of the US Labour Market." *Feminist Economics*, forthcoming in 2013, 19(2).

Banco de Mexico (2008). Database at www.banxico.org.mx (accessed on February 25, 2008), Dirección General de Plantación, SHCP.

Barr, A. and Clark, D. (2007). "A Multidimensional Analysis of Adaptation in a Developing Country Context." Working Paper 279, Centre for the Study of African Economies, The Berkeley Electronic Press.

Basu, K. (1987). "Achievements, Capabilities and the Concept of Well-Being—A Review of Commodities and Capabilities by Amartya Sen." *Social Choice and Welfare* 4: 69–76.

Basu, K. (2006). "Gender and Say: A Model of Household Behaviour with Endogenously Determined Balance of Power." *The Economic Journal* 116(April): 558–580.

Becker, G. S. (1976). *The Economic Approach to Human Behavior.* Chicago: University of Chicago Press.

Belfield, C. and Heywood, J. S. (2006). "Product Market Structure and Gender Discrimination in the United Kingdom." In J. S. Heywood and J. H. Peoples (eds), *Product Market Structure and Labor Market Discrimination.* New York: State University of New York. 39–58.

Beutelspacher, A. N., Zapata Martelo, E., and Vasquez Garcia, V. (2003). "Does Contraception Benefit Women? Structure, Agency, Well-Being in Rural Mexico." *Feminist Economics* 9: 213–238.

Blanchflower, D. and Oswald, A. (1994). *The Wage Curve*. MIT Press.

Blau, F. D., Ferber, M. F., and Winkler, A. E. (2002). *The Economics of Women, Men, and Work*, 4th edn. New Jersey: Prentice Hall.

Bonke, J. (2005). "Session 7—Well-Being and Deprivation: Subjective and Objective Measures Utilizing Time-Use Data." Paper presented in a session of a Levy Institute Conference on Time Use and Economic Well-Being, October 2005. In *The Levy Economics Institute of Bard College Report* 16(1).

Borjas, G. J. (2005). *Labor Economics*, 3rd edn. Harvard University: McGraw-Hill.

Bowden, C. (2010). *Murder City: Ciudad Juárez and the Global Economy's New Killing Fields*. New York: Nation Books.

Bowles, S. (1998). "Endogenous Preferences: The Cultural Consequences of Markets and Other Economic Institutions." *Journal of Economic Literature* 36: 75–111.

Brooks-Gunn, J., Fuligni, A. S., and Berlin, L. J. (eds) (2003). *Early Child Development in the 21st Century: Profiles of Current Research Initiatives*. London: Teachers College Press.

Burchardt, T. (2005). "Are One Man's Rags Another Man's Riches? Identifying Adaptive Expectations Using Panel Data." *Social Indicators Research* 74(1): 57–102.

Burchardt, T. and Vizard, P. (2011). "'Operationalizing' the Capability Approach as a Basis for Equality and Human Rights Monitoring in Twenty-first-century Britain." *Journal of Human Development and Capabilities* 12(1): 91–119.

Burke, M. and Peyton Young, H. (2010). "Social Norms." In Benhabib, J., Bisin, A., and Jackson, M. O. (eds), *Handbook of Social Economics*, Vol.1A. Elsevier Science and Technology

Carabelli, A. M. and Cedrini, M. A. (2009). "The Economic Problem of Happiness: Keynes on Happiness and Economics." *Working Paper 123*, SEMEQ Department, Faculty of Economics, University of Eastern Piedmont, Italy.

Carlin, W. and Soskice, D. (2005). *Macroeconomics and the Wage Bargain: A Modern Approach to Employment, Inflation, and the Exchange Rate*, 2nd edn. Oxford: Oxford University Press.

Carter, T. J. (2005). "Money and Efficiency Wages: The Neglected Effect of Employment on Efficiency." *The Journal of Socio-Economics* 34: 199–209.

Casey, T. and Dustmann, C. (2010). "Immigrants' Identity, Economic Outcomes and the Transmission of Identity across Generations." *The Economic Journal* 120 (February), F31–F51.

Catanzarite, L. M. and Strober, M. H. (1993). "The Gender Recomposition of the Maquiladora Workforce in Ciudad Juarez." *Industrial Relations* 32(1): 133–147.

Chang, C. and Wang, Y. (1996). "Human Capital Investment under Asymmetric Information: The Pigovian Conjecture Revisited." *Journal of Labor Economics* 14(3): 505–519.

Chang, H -J. (2007). *Bad Samaritans—Rich Nations, Poor Policies, and the Threat to the Developing World.* London: Random House Business Books.

Chang, W -C. (2011). "Identity, Gender, and Subjective Well-Being." *Review of Social Economy* 69(1): 97–121.

Chant, S. with Craske, N. (2003). *Gender in Latin America.* London: Latin American Bureau.

Charles, A. (2011a). "Fairness and Wages in Mexico's Maquiladora Industry: An Empirical Analysis on Labour Demand and the Gender Wage Gap." *Review of Social Economy* 69(1): 1–28

Charles, A. (2011b). "The Great Recession and Ethnic Inequality in the US Labour Force." *History of Economic Ideas* 19(2): 163–176.

Charles, A. (2012). "Social Conventions and Fairness." In Philip Arestis and Malcolm Sawyer (eds), *The Elgar Companion to Radical Political Economy,* forthcoming. Cheltenham, UK: Edward Elgar.

Clark, D. A. (2003). "Concepts and Perceptions of Human Well-Being: Some Evidence from South Africa." *Oxford Development Studies* 31(2): 173–196.

Clark, D. A. (2006). "Capability Approach." In Clark, D. A (ed.), *The Elgar Companion to Development Studies.* Cheltenham: Edward Elgar.

Clark, D. A. (2009). "Adaptation, Poverty and Well-Being: Some Issues and Observations with Special reference to the Capability Approach and Development Studies." *Journal of Human Development and Capabilities* 10(1): 21–42.

Clark, D. A. and Qizilbash, M. (2005). "The Capability Approach and Fuzzy Poverty Measures: An Application to the South African Context." *Social Indicators Research* 74: 103–139.

Comim, F. (2005). "Capabilities and Happiness: Potential Synergies." *Review of Social Economy* 63(2): 161–176.

Comim, F. (2008). "Social Capital Theory and Capabilities." In D. Castiglione, J. W. van Deth, and G. Wolleb (eds), *The Handbook of Social Capital.* Oxford: Oxford University Press.

Conlisk, J. (1996). "Why Bounded Rationality?" *Journal of Economic Literature* 34: 669–700.

Current Population Survey (CPS) (2010). Bureau of Labour Statistics, US Census Bureau (accessed on April 30, 2010 at www.bls.gov /news.release/wkyeng.nr0.htm).

Danthine, J -P. and Kurmann, A. (2006). "Efficiency Wages Revisited: The Internal Reference Perspective." *Economics Letters* 90: 278–284.

Darity, W. A. Jr. (2005). "Stratification Economics: The Role of Intergroup Inequality." *Journal of Economics and Finance* 29(2): 144–153.

Dasgupta, P. (2001). *Human Well-Being and the Natural Environment.* New York: Oxford University Press.

Dauphin, A., El Lahga, A -R., Fortin, B., and Lacroix, G. (2011). "Are Children Decision-Makers within the Household?" *The Economic Journal* 121(June): 871–903.

Davidson, R. J. (2000). "Affective Style, Psychopathology and Resilience: Brain Mechanisms and Plasticity." *American Psychology* 55: 1196–1214.

Davis, J. B. (2011). *Individuals and Identity in Economics.* Cambridge: Cambridge University Press

DeMartino, G. F. (2010). *The Economist's Oath: On the Need for and Content of Professional Economic Ethics.* Oxford: Oxford University Press

Diener, E. and Suh, E. M. (eds) (2000). *Culture and Subjective Well-Being.* Cambridge, MA: MIT Press.

Djankov, S., Glaeser, E., La Porta, R., Lopez-de-Silanes, F., and Shleifer, A. (2003). "The New Comparative Economics." *Journal of Comparative Economics* 31: 595–619.

Djebbari, H. (2005). "The Impact on Nutrition of the Intrahousehold Distribution of Power." Discussion Paper (1701), Institute for the Study of Labor (IZA), Bonn.

Duflo, E. and Udry, C. (2004). "Intrahousehold Resource Allocation in Cote d'Ivoire: Social Norms, Separate Accounts, and Consumption Choices." NBER Working Paper Series, Working Paper 10498, Cambridge, MA.

Dunlop, J. T. (1938). "The Movement of Real and Money Wage Rates." *The Economic Journal* 48(191): 413–434.

Dussel, E. P. (2001). "Has NAFTA Contributed to Economic Development in Mexico?" *Perspective.* Center for International Finance and Development, University of Iowa.

Easterlin, R. A. (2005). "Building a Better Theory of Well-Being." In L. Bruni and P. L. Porta (eds), *Economics and Happiness: Framing the Analysis.* Oxford: Oxford University Press.

Ehrenberg, R. G. and Smith, R. S. (2006). *Modern Labor Economics: Theory and Public Policy*, 9th edn. Pearson International: Cornell University.

Elsby, M. W., Hobijn, B., and Sahin, A. (2010). "The Labor Market in the Great Recession." *NBER Working Paper Series*, Working Paper 15979 (May).

Elson, D. and Pearson, R. (1981). "'Nimble Fingers Make Cheap Workers': An Analysis of Women's Employment in Third World Export Manufacturing." *Feminist Review* 7: 87–107.

Elster, J. (1989). "Social Norms and Economic Theory." *The Journal of Economic Perspectives* 3(4): 99–117.

Erikson, E. (1968). *Identity: Youth and Crisis*. New York: Norton.

Fehr, E. and Fischbacher, U. (2002). "Why Social Preferences Matter— The Impact of Non-Selfish Motives on Competition, Cooperation and Incentives." *The Economic Journal* 112 (March): C1–C33.

Fehr, E. and Schmidt, K. M. (1999). "A Theory of Fairness, Competition, and Cooperation." *The Quarterly Journal of Economics* 114(3): 817–868.

Fehr, E., Gatcher, S., Kirchler, E., and Welchbold, A. (1998). "When Social Norms Overpower Competition: Gift Exchange in Experimental Labor Markets." *Journal of Labor Economics* 16(2): 324–351.

Fehr, E., Goette, L., and Zehnder, C. (2009). "A Behavioral Account of the Labor Market: The Role of Fairness Concerns." *Annual Reviews Economics 2009*, 1: 355–384.

Fleck, S. (2001). "A Gender Perspective on Maquila Employment and Wages in Mexico." In E. Katz and M. Correia (eds), *The Economics of Gender in Mexico: Work, Family, State and Market*. Washington: The World Bank.

Folbre, N. (1986). "Hearts and Spades: Paradigms of Household Economics." *World Development* 14(2): 245–255.

Frey, B. (2008). *Happiness. A Revolution in Economics*. Cambridge: MIT Press.

Frey, B. S. and Stutzer, A. (2002). *Happiness and Economics: How the Economy and Institutions Affect Well-Being*. Princeton: Princeton University Press.

Fussell, E. (2000). "Making Labor Flexible: The Recomposition of Tijuana's Maquiladora Female Labor Force." *Feminist Economics* 6(3): 59–79.

Galbraith, J. K. (2011). "Inequality and Economic and Political Change: A Comparative Perspective." *Cambridge Journal of Regions, Economy and Society* 4(1): 13–28.

Gasper, D. and Van Staveren, I. (2003). "Development as Freedom—and as What Else?" *Feminist Economics* 9(2/3): 137–161.

Ghiara, G. (1999). "Impact of Trade Liberalisation on Female Wages in Mexico." *Development Policy Review* 17(2): 171–190.

Goette, L., Huffman, D., and Meier, S. (2006). "The Impact of Group Membership on Cooperation and Norm Enforcement: Evidence Using Random Assignment to Real Social Groups." *The American Economic Review* 96(2): 212–216.

Gruben, W. (2001). "Was NAFTA Behind Mexico's High Maquiladora Growth?" *Economic and Financial Review*, FRB of Dallas (Third Quarter). 11–21.

Hamermesh, D. S. (1993). *Labor Demand.* Princeton: Princeton University Press.

Handa, S. (1994). "Gender, Headship and Intrahousehold Resource Allocation." *World Development* 22: 1535–1547.

Hare, R. M. (1976). "Ethical Theory and Utilitarianism." In H. D. Lewis (ed.), *Contemporary British Philosophy.* London: Allen and Unwin. Reprinted in A. K. Sen and B. Williams (1982), 23–38.

Harsanyi, J. C. (1977). "Morality and the Theory of Rational Behaviour." *Social Research* 44(4). Reprinted in A. K. Sen and B. Williams (1982), 39–62.

Herzog, T. (1996). *Research Methods in the Social Sciences.* New Jersey: Prentice-Hall.

Huck, S., Kirchsteiger, G., and Oechssler, J. (2005). "Learning to Like What You Have. Explaining the Endowment Effect." *The Economic Journal* 115(505): 689–702.

Hume, D. (1777). *Enquiries Concerning Human Understanding and Concerning the Principles of Morals.* In L. A. Selby-Brigge (ed.), revised edn. (1975). Oxford: Clarendon Press.

INEGI (2011). Instituto Nacional de Estadística Geografía e Informática, www.inegi.gob.mx (accessed sporadically from 2004 to 2011).

Infomat (2000). *Apparel and Textile Production.* Research Report, www.infomat.com/infre0000249.html (accessed on October 01, 2008).

Infomat (2006). *Textile Industry in Mexico.* Research Report, www.infomat.com/infre0000230.html (accessed on October 01, 2008).

Inglehart, R. and Klingemann, H -D. (2000). "Genes, Culture, Democracy and Happiness." In E. Diener and E. M. Suh (eds), *Culture and Subjective Well-Being.* Cambridge, MA: MIT Press.

Iversen, V. (2003). "Intra-Household Inequality: A Challenge for the Capability Approach?" *Feminist Economics* 9(2/3): 93–115.

Kahneman, D. and Tversky, A (1986). "Rational Choice and the Framing of Decisions." *The Journal of Business* 59(4): S251–S278.

Kahneman, D. (2000). "Experienced Utility and Objective Happiness—A Moment-Based Approach." In D. Kahneman and A. Tversky (eds), *Choices, Values and Frame*. Cambridge: Cambridge University Press.

Kahneman, D., Diener, E., and Schwarz, N. (eds) (1999). *Well-Being: The Foundations of Hedonic Psychology*. New York: Russell Sage Foundation.

Kahneman, D., Knetsch, J. L., and Thaler, R. (1986). "Fairness as a Constraint on Profit Seeking: Entitlements in the Market." *The American Economic Review* 76(4): 728–741.

Kahneman, D. and Krueger, A. B. (2006). "Developments in the Measurement of Subjective Well-Being." *Journal of Economic Perspectives* 20(1): 3–24.

Kahneman, D. and Tversky, A. (1979). "Prospect Theory: An Analysis of Decision under Risk." *Econometrica* 47: 263–291.

Kahneman, D. and Tversky, A. (2000). *Choices, Values and Frame*. Cambridge: Cambridge University Press.

Katz, E. (1997). "The Intra-Household Economics of Voice and Exit." *Feminist Economics* 3(3): 25–46.

Katz, E. G. and Correia, M. C. (2001). *The Economics of Gender in Mexico: Work, Family, State and Market*. Washington, DC: The World Bank, Directions in Development.

Kennedy, E. and Peters, P. (1992). "Household Food Security and Child Nutrition: The Interaction of Income and Gender of Household Head." *World Development* 20: 1077–1085.

KILM (2005). *Key Indicators of the Labour Market*, 4th edn. Geneva: International Labour Office, Bureau of Publications.

Klasen, S. (1997). "Poverty, Inequality and Deprivation in South Africa: An Analysis of the 1993 SALDRU Survey." *Social Indicators Research* 41: 51–94.

Koch, P. O. (2006). *The Aztecs, the Conquistadors, and the Making of Mexican Culture*. Jefferson, NC: McFarland.

Koford, K. J. and J. B. Miller (1991). *Social Norms and Economic Institutions*. Michigan: The University of Michigan Press.

Lancaster, G., Maitra, P., and Ray, R. (2006). "Endogenous Intra-Household Balance of Power and Its Impact on Expenditure Patterns: Evidence from India." *Economica* 73: 435–460

Layard, R. (2003). "Happiness: Has Social Science a Clue?" *Lionel Robbins Memorial Lectures* (3–5 March), London School of Economics (accessed on August 08, 2007 at http://cep.lse.ac.uk /events/lectures/layard/RL030303.pdf).

Lewis, J. and Giullari, S. (2005). "The Adult Worker Model Family, Gender Equality and Care: The Search for New Policy Principles and the Possibilities and Problems of a Capabilities Approach." *Economy and Society* 34(1): 76–104.

López, J. G., Sánchez A. V., Contreras-Cristan A., and Chang, M. (2008). "Money Wages in Mexico: A Tale of Two Industries." *Investigacióneconómica* 67(266): 14–29.

March, J. G. (1988). "Variable Risk Preferences and Adaptive Aspirations." *Journal of Economic Behavior and Organization* 9(1): 5–24.

Marjit, S., Ghosh, S., and Biswas, A. (2007). "Informality, Corruption and Trade Reform." *European Journal of Political Economy* 23(3): 777–789.

Mason, P. L. (1995). "Race, Competition and Differential Wages." *Cambridge Journal of Economics* 19(4): 545–567.

Michie, J. (1987). *Wages in the Business Cycle: An Empirical and Methodological Analysis.* London: Pinter.

Mill, J. S. (1956). *A System of Logic Ratiocinative and Inductive being a Connected View of the Principles of Evidence and the Methods of Scientific Investigation,* 8th ed. London: Longmans, Green and Co.

Mollick, A. V. and Wvalle-Vazquez (2006). "Chinese Competition and Its Effects on Mexican Maquiladoras." *Journal of Comparative Economics* 34: 130–145.

Neumann, J. von and Morgenstern, O. (1944). *Theory of Games and Economic Behavior.* Princeton, NJ: Princeton University Press.

Nijkamp, P. and Poot, J. (2005). "The Last Word on the Wage Curve?" *Journal of Economic Surveys* 19(3): 421–450.

Nussbaum, M. C. (2000). *Women and Human Development: The Capabilities Approach.* Cambridge: Cambridge University Press.

Nussbaum, M. C. (2003). "Capabilities as Fundamental Entitlements: Sen and Social Justice." *Feminist Economics* 9(2/3): 33–59.

Nussbaum, M. C. (2005). "Mill between Aristotle and Bentham." In L. Bruni and P. L. Porta (eds), *Economics and Happiness: Framing the Analysis.* Oxford: Oxford University Press. 170–183.

Nussbaum, M. C. (2006). *Frontiers of Justice: Disability, Nationality, Species Membership.* Belknap: Harvard University Press.

Nussbaum, M. C. (2011). "Capabilities, Entitlements, Rights: Supplementation and Critique." *Journal of Human Development and Capabilities* 12(1): 23–37.

Palma, G. (2003). "Trade Liberalization in Mexico: Its Impact on Growth, Employment and Wages." *Employment Paper* (2003/55). Geneva: International Labour Office.

Paz, O. (1985). *The Labyrinth of Solitude and Other Writings.* New York: Grove Press

Pollin, R. and Zhu, A. (2006). "Inflation and Economic Growth: A Cross-Country Nonlinear Analysis." *Journal of Post Keynesian Economics* 28(4): 593–614.

Qizilbash, M. (1997). "Policy Arena—A Weakness of the Capability Approach with Respect to Gender Justice." *Journal of International Development* 9(2): 251–262.

Rabin, M. (1993). "Incorporating Fairness into Game Theory and Economics." *The American Economic Review* 83(5): 1281–1302

Rabin, M. (2000). "Diminishing Marginal Utility of Wealth Cannot Explain Risk Aversion." In D. Kahneman and A. Tversky (eds), *Choices, Values and Frames.* Cambridge: Cambridge University Press.

Rader, T. (1980). "The Second Theorem of Welfare Economics When Utilities Are Interdependent." *Journal of Economic Theory* 23: 420–424.

Rangel, M. A. (2006). "Alimony Rights and Intra-Household Allocation of Resources: Evidence from Brazil." *The Economic Journal* 116(July): 627–658.

Ravallion, M. (2010). "Troubling Tradeoffs in the Human Development Index." Policy Research Working Paper 5484. Washington, DC: The World Bank.

Rayo, L. and Becker, G. S. (2005). "On the Foundations of Happiness." University of Chicago: Mimeo.

Robertson, R. (2005). "Has NAFTA Increased Labor Market Integration between the United States and Mexico?" *The World Bank Economic Review* 19(3): 425–448.

Robeyns, I. (2005a). "Selecting Capabilities for Quality of Life Measurement." *Social Indicators Research* 74(1): 191–215.

Robeyns, I. (2005b). "The Capability Approach: A Theoretical Survey." *Journal of Human Development* 6(1): 97–117.

Robinson, J. (1969). *The Economics of Imperfect Competition*, 2nd edn. London: Macmillan.

Rodgers, Y. (2004). "Labor Laws and Policies in Asia: Impacts, Tradeoffs, and Implications for the End of the Multi-Fiber Agreement." Rutgers University: Mimeo.

Rodgers, Y. and Nataraj, S. (1999). "Labor Market Flexibility in East Asia: Lessons from Taiwan." *Economic Development and Cultural Change* 48 (1): 51–69.

Ruwanpura, K. N. (2008). "Multiple Identities, Multiple-Discrimination: A Critical Review." *Feminist Economics* 14(3): 77–105.

Şahin, A., B. Hobijn, and J. Song (2010). "The Unemployment Gender Gap during the 2007 Recession." *Current Issues in Economics and Finance*, Federal Reserve Bank of New York. 16(2): 1–7.

Sargent, J. and Matthews, L. (2004). "What Happens when Relative Costs Increase in Export Processing Zones? Technology, Regional Production Networks, and Mexico's Maquiladoras." *World Development* 32(12): 2015–2030.

Sawyer, M. C. (2002). "The NAIRU, Aggregate Demand, and Investment." *Metroeconomica* 53(1): 66–94.

Sawyer, M. C. and Spencer, D. (2010). "Labour Supply, Employment and Unemployment in Macroeconomics: A Critical Appraisal." *Review of Political Economy* 22(2): 263–279.

Schönberg, U. (2007). "Testing for Asymmetric Employer Learning." *Journal of Labor Economics* 25(4): 651–691.

Seguino, S. (2000). "Gender Inequality and Economic Growth: A Cross-Country Analysis." *World Development* 28(7): 1211–1230.

Sen, A. K. (1981). *Poverty and Famines: An Essay on Entitlement and Deprivation.* Oxford: Oxford University Press.

Sen, A. K. (1982). *Choice, Welfare and Measurement.* Oxford: Oxford University Press.

Sen, A. K. (1985). *Commodities and Capabilities.* Oxford: Elsevier Science Publishers.

Sen, A. K. (1987). *The Standard of Living.* The Tanner Lectures, Clare Hall, Cambridge, UK.

Sen, A. K. (1990). "Gender and Cooperative Conflicts." In Tinker, I. (ed), *Persistent Inequalities: Women and World Development.* Oxford: Oxford University Press.

Sen, A. K. (1992). *Inequality Reexamined.* Oxford: Oxford University Press.

Sen, A. K. (1999). *Development as Freedom.* Oxford: Oxford University Press.

Sen, A. K. (2005). "Identity as a Dividing Force." Open Lecture given in Robinson College, June 2005, Cambridge, UK.

Shafir, E, Diamond, P., and Tversky, A. (1997). "Money Illusion." *The Quarterly Journal of Economics* 112(2): 341–374.

Shapiro, C. and Stiglitz, J. (1984). "Equilibrium Unemployment as a Worker Discipline Device." *American Economic Review* 74: 433–444.

Shoham, Y. and M. Tennenholtz (1997). "On the Emergence of Social Conventions: Modeling, Analysis and Simulations." *Artificial Intelligence* 94: 139–166

Singer, P. (2002). "A Response to Nussbaum"; Reply to Martha Nussbaum, "Justice for Non-Human Animals." The Tanner Lectures on Human Values, November 13, http://www.utilitarian.net/singer/by/20021113.htm (accessed on December 15, 2007).

Şirin Saracoğlu, D. (2007). "The Informal Sector and Tax on Employment: A Dynamic General Equilibrium Investigation." *Journal of Economic Dynamics and Control* 32(2): 529–549.

Sklair, L. (1989). *Assembling for Development: The Maquila Industry in Mexico and the United States.* Boston: Unwin Hyman.

Skott, P. (2005). "Fairness as a Source of Hysteresis in Employment and Relative Wages." *Journal of Economic Behavior and Organization* 57: 305–331.

Smith, V. L. (2003). "Constructivist and Ecological Rationality in Economics." *American Economic Review* 93(3): 465–508.

Steckel, R. H. (2008). "Biological Measures of the Standard of Living." *Journal of Economic Perspectives* 22(1): 129–152.

Steiner, H. (1987). "Entitlements." In J. Eatwell, M. Milgate, and P. Newman (eds), *The New Palgrave: A Dictionary of Economics.* London: Macmillan.

Stewart, J. (2008). "African Studies and Economics: In Search of a New Progressive Partnership." *Journal of Black Studies* 38(5): 795–805.

Sunder, S. (2009). "Book Review: Rationality in Economics: Constructivist and Ecological Forms, by Vernon L. Smith." *Journal of Economic Psychology* 30: 107–110.

Tajfel, H. and Turner, J. (1986). "The Social Identity Theory of Intergroup Behavior." In S. Worchel and W. Austin (eds), *Psychology of Intergroup Relations*, 7–24. Chicago: Nelson-Hall.

Tcherneva, P. R. (2006). "Chartalism and the Tax-Driven Approach to Money." In P. Arestis and M. C. Sawyer (eds), *A Handbook of Alternative Monetary Economics.* Cheltenham, UK and Northampton, MA: Edward Elgar.

Tymoigne, E. and Wray, L. R. (2006). "Money: An Alternative Story." In P. Arestis and M. C. Sawyer (eds), *A Handbook of Alternative*

Monetary Economics. Cheltenham, UK and Northampton, MA: Edward Elgar.

UNDP (2005). *Human Development Report 2005*. http://hdr.undp .org. (accessed on June 6, 2007).

UNICEF (2007). *Innocenti Report Card 7*. Florence: Innocenti Research Centre.

Urry, H., Nitschke, J., Dolski, I., Jackson, D., Dalton, K., Mueler, C., Rosenkranz, M., Ryff, C., Singer, B., and Davidson, R. (2004). "Making a Life Worth Living." *Psychological Science* 15(6): 367–372.

US Census Bureau (2006). "Hispanic in the United States." Ethnicity and Ancestry Branch Population Division.

Uyan-Semerci, P. (2007). "A Relational Account of Nussbaum's List of Capabilities." *Journal of Human Development* 8(2): 203–221.

Van Staveren, I. (2008). "Capabilities and Well-Being." In J. B. Davis and W. Dolfsma (eds), *The Elgar Companion to Social Economics*. Cheltenham: Edward Elgar Publishing.

Veenhoven, R. (2000). "Freedom and Happiness: A Comparative Study in Forty-Four Nations in the Early 1990s." In E. Diener and E. M. Suh (eds), *Culture and Subjective Well-Being*. Cambridge, MA: MIT Press.

Weber, M. (1958). "The Protestant Ethic and the Spirit of Capitalism." In M. Lambek (ed.), *A Reader in the Anthropology of Religion*. Malden, MA: Blackwell Publishers.

Welch, C. (1987). "Utilitarianism." In J. Eatwell, M. Milgate, and P. Newman (eds), *The New Palgrave: A Dictionary of Economics*. London: Macmillan.

Williams, R. M. (1993). "Race, Deconstruction, and the Emergent Agenda of Feminist Economic Theory." In M. A. Ferber and J. A. Nelson (eds), *Beyond Economic Man. Feminist Theory and Economics*. Chicago: University of Chicago Press.

Woersdorfer, J. S. (2010). "When Do Social Norms Replace Status-Seeking Consumption? An Application to the Consumption of Cleanliness." *Metroeconomica* 61(1): 35–67.

Wolf, E. R. (1958). "The Virgin of Guadalupe: A Mexican National Symbol." In M. Lambek (ed.) (2002), *A Reader in the Anthropology of Religion*. Malden, MA: Blackwell Publishers.

World Bank (2009). *World Development Indicators*. The World Bank Group, http://www.worldbank.org (accessed on January 23, 2009).

World Values Survey (2006). "Mexico Survey 2000." European and World Values Surveys Four-Wave Integrated Data File, 1981–2004, v.20060423, The European Values Study Foundation and World Values Survey Association.

Zepeda, E. (2008). "Latin America's Progress on Gender Equality: Poor Women Workers Are Still Left Behind." International Poverty Centre, UNDP, One-pager (49).

Index

absenteeism, 94
access, 2–7, 10–11, 13, 19, 29,
  32–4, 47, 49, 53–5, 57–8,
  60–1, 66–75, 79, 85–7, 90,
  102, 117, 123–4, 126, 129–32
achieved functioning, 3, 12, 28, 53,
  55–9, 64–72, 77–8, 127
act-utilitarianism, 22
adaptation, 10, 13, 16–17, 21,
  24–32, 35, 77, 81–6, 90, 100,
  116, 129, 159
  hedonic, 25, 29, 82–4
  and interdependence, 10, 13, 16,
    26, 28–32, 77, 81, 85–6,
    90, 100
  and preferences, 17, 21, 26, 29,
    81–6, 100, 129, 159
African Americans, 46, 49, 70–1
age, 4, 39, 41–3, 50, 68, 73, 98,
  113–14, 117, 120, 121, 124,
  127, 137, 144, 148, 150
Ahnert, L., 73
Akerlof, G. A., 41–2, 48, 94, 99,
  136–8, 141–2, 165n6
Alchian, A. A., 140
Algan, Y., 46
Alkire, S., 15
Altman, M., 20, 51
American Economic Association, 99
American Human Development
  Project (2010), 70
Anand, P., 56
Aquinas, T., 110

Arestis, P., 49, 165n11
Aristotle, 9–10, 14, 21
Asian identity, 46, 87–8
asymmetric information, 46, 95,
  100, 137
Aztecs, 107–9

Baja California, 112–15
Barr, A., 83
Basu, K., 16, 74, 77
Becker, G. S., 24–5, 31, 73, 163n6
Belarus, 83
Bentham, 16, 21, 24, 26
Beutelspacher, A. N., 51, 162n9
Blanchflower, D., 93, 123
Bonke, J., 85–6
bounded rationality, 66, 101, 163n7
Bowles, S., 81–2
Brazil, 6
British Airways, 50
British Household Panel Survey, 84
Brooks-Gunn, J., 73
Buddhism, 17, 44–5
Burchardt, T., 12, 16, 20, 29, 84–5

CA, *See* Capability Approach
Cambodia, 88
Canada, 135
capabilities, 12–23, 28–33, 38, 41,
  44, 53–6, 59, 61–6, 68–73, 75,
  78–80, 87, 92, 101, 137–43,
  155, 158–60
  in context, 17–21

capabilities—*Continued*
  and income, 20
  "social capabilities," 31, 64–5, 79
  subjective and objective, 12–17
  ten central human, 13–14
Capability Approach (CA), 1–7,
    9–22, 29–30, 32, 44, 60–3,
    68, 161n1,2
  and causal link, 61–2
  in context, 17–21
  defined, 1–7
  and subjectivity, 9–17
Carabelli, A. M., 24, 162n2
Carlin, W., 94
Carter, T. J., 90
Casey, T., 42, 46
caste, 4, 43, 50, 74
Catanzarite, L. M., 143–4
Catholics, 51, 106
Caucasian Americans, 46, 48–9, 70
Cedrini M. A., 24, 162n2
Chang, C., 43, 46, 164n3
Chant, S., 148
Charles, A., 136, 141, 145–6, 148,
    162n3,5
Chiapas, 51, 135
Chihuahua, 112–16
children, 19–21, 51, 59, 66–74, 79,
    84, 98, 103, 108, 117–18,
    128, 148
  care of, 68–73
  and well-being, 19–20
Chile, 6
China, 5, 33, 87, 136, 141, 143–4,
    149, 155
choice, 3–4, 11–12, 16, 18, 20, 23,
    26, 28, 30–2, 34–7, 40, 44,
    50–1, 54–7, 66, 71–4, 77–8,
    80, 84, 93–6, 102, 136–7,
    159–60, 163n7
Christianity, 17, 33, 45, 105, 109–10
Ciudad Juarez (Chihuahua),
    112–16, 133, 143–4
Clark, D. A., 10–11, 15, 18–20, 29,
    62, 83, 127

Coahuila de Zaragoza, 112–13, 115
Coatlicue, 108
Coca-Cola Company, 82
colonization, 5, 109
Comim, F., 9, 16, 31, 65, 79
commodities, 3, 6, 10–12, 25,
    32, 45, 55–72, 75, 78–80,
    83, 85–6, 98, 126–7, 133,
    158–9
  and endowments, 64–9
common interest, 36–7, 50
competition, 5, 25, 46, 87–8,
    92–100, 103, 136, 141, 143–4,
    149, 155, 164n14
Conlisk, J., 101, 163n7
conquistadores, 107
consumption bundles, 2, 53, 80
context, 2–3, 11–18, 21–2,
    29, 34–5
Correia, M., 120
Cortes, Hernán, 108

Dasgupta, P., 9
Dauphin, A., 68
Davidson, R. J., 27
Davis, J. B., 16, 41, 45, 162n5,6
death of a close relative, 25
decision-making, 29, 34, 40, 46,
    68–9, 73–4, 100–1, 137, 142
democracy, 12–15
Demsetz, H., 140
developmental theory, 42
diabetes, 116–17, 132, 158
Diener, E., 26
discrimination, 46, 50, 59, 101,
    120, 137, 149–50
Djankov, S., 99
Djebbari, H., 127
Dona Marina, 109–10
Dussel, E. P., 135–6
Dustmann, C., 42, 46

Easterlin, Richard, 24–6, 29, 69,
    83, 163n6
ecological rationality, 29, 40

economic entitlements, 4, 6–7, 53,
58–61, 64, 68, 72, 77–81, 86,
88, 91–103, 124, 132,
135–55, 159
consequences on, 92–103
insider-outside model, 137–43
and male share of
employment, 143–9
and *Maquiladora* labor market,
143–51
and social norms, 151–6
and trends in the gender wage
gap, 149–51
Ejército Zapatista de Liberación
Nacional (EZLN), 135
electroencephalograph, 27
Elsby, M. W., 47–8
Elson, D., 138
E-mapping, 1–4, 7, 10–11, 53–75,
77–103
and capabilities, 55–64
and endowment, 64–9
external shocks on, 77–103
identity, *See* identity E-mapping
and *maquiladora* workers, 87–91
"empirical philosophy," 18–19
employment, 6, 19, 37–9, 46–9, 68,
75, 79, 85–102, 106, 120–5,
135–9, 143–51, 155, 159,
164n14
full-time, 48–9
informal, 91, 93–4, 99, 114,
120–5, 144, 151, 155
limited access to, 6
male share of, 143–9
part-time, 48–9
underemployment, 122–4
See also unemployment
"endogenous preferences," 81–2
"endowment effect," 28–9, 81
endowments, 2–4, 28–9, 51–2, 53,
55, 64–73, 78, 80–2, 85–6,
128, 137
and commodities, 64–9
individual, 65–7, 78, 81

and *maquiladora*
households, 128
entitlement failures, 1–4, 7, 15, 27,
32, 53–6, 63, 65, 68–71, 75,
80, 160
entitlements
economic, *See* economic
entitlements
failures, *See* entitlement failures
social, *See* social entitlements
entrepreneurs, 99–101
EPZ, *See* Export-Processing Zone
"equality," 44
equivalence, 99
Erikson, E., 42
ethics, 21–2, 35, 44, 64, 162n8
ethnicity, 4, 14, 39, 41–3,
47–9, 159
eudaimonia, 9–10, 21
Evangelists, 106
evolution, 24–5, 34–5, 40
exchange-entitlement mapping,
See E-mapping
exchange rate, 5, 53, 89–90, 95–8
Exogeneity Wald tests, 154
expectations, problem of, 99–101
experienced utility, 26–8, 42–3
Export-Processing Zone
(EPZ), 5, 78
external shocks, 77–103, 111–19
on E-mapping, 77–103
on *maquiladora*'s E-mapping,
87–91
on Mexican identity, 111–19
externalities, 45, 95
EZLN, *See* Ejército Zapatista de
Liberación Nacional

fairness, 35–40, 45, 49, 60, 63, 68,
80–1, 90, 94, 97, 99–100, 120,
125, 141–2, 149, 151, 155
"fair-wage constraints," 38, 49, 94,
141–2, 149, 155
family, 19, 25–6, 30–1, 51, 59, 66,
72, 91, 109, 117, 127, 163n7

Fehr, E., 36–8, 46, 162n2
Fischbacher, U., 37, 162n2
Fleck, S., 144
France, 44, 46
"fraternity," 44
freedom, 3, 6–7, 10, 12, 14–16,
    20–1, 30–1, 44, 54–7, 69,
    77, 80
free-rider behavior, 34
Frey, B., 26, 89
functioning, 3, 6, 10–12, 16–17,
    20, 28–31, 34, 51–2, 53–75,
    77–81, 86, 88, 103, 114, 116,
    126–8, 129–32, 137, 158–60
  achieved, 3, 12, 28, 53, 55–9,
    64–72, 77–8, 127
  and capability, 55–6
  potential, 3, 6, 11–12, 28, 31,
    53–6, 58, 63–4, 66, 68, 70–2,
    77–9, 86, 103, 126, 159–60
  and time, See time
Fussell, E., 144
"fuzzy poverty measures," 19

Galbraith, James, 157
game theory, 34–6
Gasper, D., 16
GEM, See gender empowerment
    measure
gender, 4, 6, 36, 39, 41, 43–4,
    49–51, 60, 66, 68, 70, 74–5,
    85–6, 90, 93, 95, 102–3,
    105–6, 109–12, 115–16,
    117–32, 136–7, 139, 141–3,
    145, 148–51, 155, 157, 162n9
gender empowerment measure
    (GEM), 125
gender identity, 119–32
  in the labor market, 119–26
  within *Maquiladora* households,
    126–32
gender norms, 66
Germany, 46–7
Ghiara, G., 120–1, 124
gift-exchange relationship, 37–8, 94

Giullari, S., 68
"Glass Ceiling effect," 49, 165n7
Glitz, A., 46
global economic crisis (2007–2010),
    47–8, 141, 144, 149, 155, 157
globalization, 78–9
Goette, L., 44
Granger causality tests, 154
Great Britain, 10, 44, 50, 68–9, 84
"Great Recession" (2007–2010),
    47–8, 141, 144, 149, 155, 157
Greenpeace, 22
group identity, 7, 35, 37, 40–4, 55,
    77, 80, 83, 160
Gruben, W., 87

HA, *See* Happiness Approach
Hamermesh, D. S., 97
happiness, 1, 9, 21, 23–31, 54, 69,
    83–5, 89, 162n2
Happiness Approach (HA), 1–2, 7,
    9–10, 21–32
  defined, 21–8
  and individual happiness, 23–8
  and interdependence and
    adaptation, 28–32
  and utilitarian tradition, 21–3
Harsanyi, J. C., 22–3
HDI, *See* Human Development
    Index
health, 12–13, 19–21, 25–7, 51,
    56, 58, 65–6, 69, 83, 88, 106,
    116–21, 127, 132
hedonic adaptation, 25, 29, 82–4
hedonistic utilitarianism, 23
Hispanic identity, 46, 49, 109, 133,
    165n11
homicides, 106, 114–16
household economy, 5–6, 19–21,
    29, 33, 45–6, 51, 60, 63,
    65–8, 71–5, 79–80, 84–91,
    98, 100–3, 105–10, 117–32,
    135–6, 141–2, 147–51, 158–9,
    163n9
Huck, S., 29–30, 81

Huffman, D., 44
Huitzilopochtli, 107–10
human capabilities, central, 13–14
  See also capabilities
Human Development Index (HDI),
  12–13, 17–18, 20, 125
Human Development Reports, 13
human rights, 14, 16, 60–1, 69
humanism, 14, 45
Hume, David, 36
hygiene, 20–1, 162n7

ideals, 1–2, 4, 7, 10, 14–15, 23, 28,
  31, 33–4, 40–6, 53–4, 59, 61,
  69–70, 75, 81, 83–4, 89, 99,
  100, 103, 105–11, 117, 126,
  129–32, 136–40, 143, 158–9
historical, 106–11
and identity, 40–5
identity/identities, 33–52
economics, 41
E-mapping, *See* identity
  E-mapping
and ideals, 40–5
and inequality, 45–52
multiple, *See* multiple identities
and norms, 33–52
perception of, 43
personal, 41–2
identity economics, 41
identity E-mapping, 7, 63, 69–75,
  79–87, 102, 159
circular relationship of, 63
external shocks on, 79–87
ILO, *See* International Labor
  Organization
immigrants, 20, 46–7, 51, 101, 111
incentives, 5–6, 36–7, 47, 137,
  140, 143
Income distribution, 49, 80, 116
individual agency, 42
individual happiness, 23–8
INEGI, *See* Instituto Nacional
  de Estadística Geografía e
  Informática

inequality, 3–4, 6, 10–14, 32, 34,
  36–9, 41, 45–52, 54, 63,
  70–1, 75, 78, 125, 132,
  157–8
and identities, 45–52
inequity aversion, 36–7
inflation, 53, 59, 84, 89–91, 98–9,
  112, 135, 147–8, 164n11
informal employment, 91, 93–4,
  99, 114, 120–5, 144,
  151, 156
insider-outsider, 41–3, 137–43
Instituto Nacional de Estadística
  Geografía e Informática
  (INEGI), 87–8, 91, 98, 113,
  115, 118–19, 120–4, 139,
  147, 151
interdependence, 3–7, 10, 13, 16,
  21, 26, 28–32
interest rate, 53
International Labor Organization
  (ILO), 123
irrational preferences, 163n7
Islam, 17, 33
Italy/Italian, 20, 51, 60–1
Iversen, V., 16

Japan, 18
Jesus Christ, 110
Jews, 20–1, 51
jobs, 4, 7, 25, 38–40, 46–9, 59–60,
  72, 75, 86, 90–1, 94, 97–8,
  100–2, 125–30, 136,
  139–44
Judaism, 45

Kahneman, Daniel, 23–4, 26–9
Katz, E., 73, 120
Keynes, 24, 54, 95, 162n2
Klasen, S., 19
Koch, P. O., 107
Koford, K. J., 35
Kranton, R. E., 41–2, 48,
  136–8, 142
Krueger, A. B., 26–7, 43

labor market, 5–6, 25, 37–9, 45–50, 60, 67–8, 85, 91, 98, 102–3, 106–7, 117–26, 127, 129–33, 137, 141–3, 149–51, 155, 165n6
Lamontagne, L., 20, 51
Lancaster, G., 74
Layard, R., 23–4, 27, 82–3
legitimacy, rule of, 60
Lewis, J., 68
"liberty," 44
life expectancy, 12, 17–18, 39
Lithuania, 83
living standards, 7, 19, 61, 82–5, 91, 100, 112, 133, 148, 150, 158
López, J. G., 91

Manning, A., 46
maquiladora industry, 4–7, 78–9, 87–91, 91, 93–8, 102, 105–6, 112, 116, 122, 124–32, 135–55
　description of, 78–9
　and economic growth, 136
　employment proportion and growth of workers, 145–7
　employment trend of line workers by gender, 139
　and endowment of resources by gender, 128
　external shocks on E-mapping, 87–91, 102
　and female homicides, 116
　and fertility rate, 98
　and gender, 6–7, 102–3, 105, 116, 126–32, 136–54
　and gender identity within households, 126–32
　and gender roles, 6–7
　gender wage gap trends, 149–51
　and Granger causality/ Exogeneity Wald tests, 154
　and identity, 105–6
　and identity utility, 142–3
　and insider-outsider model, 137–43
　and labor market trends, 143–51
　and labor supply, 98
　and male share of employment, 143–9
　and mobility, 112–13
　and nominal wages by gender, 139
　and norms and ideals, 140–2
　overview of, 4–5
　perceived optimality between spouses, 130–1
　productivity log, 152
　real wages, 152
　and simultaneity in supply and demand, 96–7
　and social norms and productivity, 151–6
　and variance decomposition, 153
　and wages, See wages and maquiladora industry
March, J. G., 29
market interactions, 2, 5–6, 14, 25, 34–50, 53, 59–64, 67–70, 73, 75, 80, 82, 85–7, 90–8, 102–3, 106–7, 117–26, 127, 129–33, 137, 141–3, 149–51, 155–60
Marx, Karl, 10
Matthews, L., 136, 143
maximization, 31, 34, 38, 45–6, 69, 74, 84, 87, 92, 94, 96–7, 100, 149, 157
　profit, 38, 45–6, 84, 87, 92, 96–7, 100, 149, 157
　utility, 31, 34, 45, 69, 84, 157
Meier, S., 44
Mexican-American war (1846–1848), 5
Mexican Consumer Price Index (CPI), 88, 112
　1969–2009, 88
Mexican Gross Domestic Product (GDP), 6, 87, 98, 112–13, 117, 119, 132, 135

Mexican Official Institution for
  Statistics, 123
Mexico, 4–6, 51, 60–1, 78, 88–9,
  98, 105–33, 135–56, 158,
  163n10
  and border states, 4–6, 78,
    105–6, 111–16, 133, 158
  and clash of cultures, 114–15
  and consumer price
    index, 88, 112
  and cultural identity and
    historical ideals, 106–11
  diet of, 60–1
  and employment rates, 120–1
  and exports, 98, 135
  and external shocks, 111–19
  and femininity, 105, 107, 110–11,
    126, 144
  and fertility, 117
  and GDP, 135
  and gender, 106–11, 117–26
  and gender identity in the labor
    market, 119–26
  and gender norms, 106–11
  and gender roles, 106–11,
    117–21
  growth rate in different aspects of
    health, 118
  growth rate in different aspects of
    household structure, 119
  growth rates in labor force
    composition per industry
    (1998–2003), 121
  and health, 116–17
  and homicides, 114–16
  and identity, 105–19
  and informal
    employment, 120–5
  and *maquiladora* industry, See
    maquiladora industry
  and masculinity, 105, 107,
    110–11, 126
  and NAFTA, *See* North American
    Free Trade Agreement
  in 1994, 105

and the peso, *See* peso
    devaluation
  and population growth, 114
  and population movements in
    border states, 112–14
  recession (1995), 105
  and U.S., 133
  and women's
    contraception, 51
Michie, J., 97
migration flows, 5, 88, 113–15,
  132–3, 158
Mill, J. S., 21–2
Miller, J. B., 35
modeling of intentions, 36
Mollick, A. V., 136
mortality rates, 5, 20, 106, 116–17,
  132, 158
MRI scans, 27
multiple identities, 2–4, 34, 39,
  41–5, 50, 79, 100, 137–8, 157,
  159–60
mythical hybrid, 74

NAFTA, *See* North American Free
    Trade Agreement
Nataraj, S., 123
negative feelings, 24, 27, 61
neoclassical theory, 95–7
New Mexico, 5
New York City, 20, 51
NGO, *See* nongovernmental
    organization
Nijkamp, P., 93
nongovernmental organization
  (NGO), 84
norms, 1–2, 4, 7, 9–10, 20–3, 28,
  30–2, 33–52, 54–6, 58–60,
  65–6, 73–5, 77–9, 90, 92, 94,
  99–102, 105–7, 110–11, 115,
  119, 124–7, 131–2, 136–43,
  149–50, 155, 158–60, 161n1,
  162n2,3
  of behavior, 33–4
  of fairness, *See* fairness

norms—*Continued*
  and identities and inequality,
    45–52
  and identity and ideal, 40–5
  origin and stability of, 35–9
  overview of, 33–52
  problem of, 99–101
  social, *See* social norms
North American Free Trade
  Agreement (NAFTA), 5–6, 79,
  87, 98, 105–6, 111–12, 114,
  117, 135, 138, 144, 155, 158
Nuevo Leon, 112–15
Nussbaum, Martha C., 1–2, 10–17,
  20–1, 24, 28–9, 31, 56, 60, 63,
  69, 161n3,1,2, 162n1, 163n4

objective capabilities, 12–17
Oswald, A., 93, 123

Paz, Octavio, 105, 110–11, 114
Pearson, R., 71, 138
peer comparison, 2, 24–8, 140, 157
perception, 10–11, 19, 26–9, 33–4,
  38–9, 42–3, 50, 65–6, 71–2,
  74–5, 78, 81, 85, 90–1, 95,
  99–102, 106, 124, 126–32,
  137, 141–2, 150, 155, 159
perfection, 33, 43
personal judgment, 61
personality, 26
personhood, 10, 16–17
peso devaluation, 5–6, 87–9, 98,
  105, 112, 114, 117, 135, 138
policy, 3–7, 15, 56–7, 63, 87,
  123, 158
Poot, J., 93
positive feelings, 24, 27
potential functioning, 3, 6, 11–12,
  28, 31, 53–6, 58, 63–4, 66,
  68, 70–2, 77–9, 86, 103, 126,
  159–60
poverty, 1, 6, 10–11, 19–20, 53, 60,
  89, 75, 162n3
preferences, 10, 13, 16–17,
  21–2, 163n7

productivity, 21, 37–8, 89, 94, 97,
  102, 136–43, 147, 149–56
  log of, 152
  per worker-hour, 152
  variance decomposition of, 153
  and wages, 151–6
professional athletes, 34
profit maximization, 34, 38, 45–6,
  84, 87, 92, 96–7, 100, 149, 157
Protestant ethic, 14, 106
psychology, 23–6, 28, 42, 81
public goods, 58, 64–6, 71, 127

Qizilbash, M., 19
Quetzalcoatl, 109

Rabin, M., 29, 35–6
Rader, T., 10, 16
Ravallion, M., 20
Rawls, John, 10
Rayo, L., 24–5, 31
real wage, 59, 67–8, 79, 88, 89,
  91–3, 95, 97–101, 122, 136,
  147–8, 151–5, 159, 164n13
recessions, 5–6, 47–8, 87, 112, 138,
  141, 144, 149, 155, 157
  and employers, 48
  and gender norms in Mexico, 141
  "Great Recession"
    (2007–present), 47–8, 141,
    144, 149, 155, 157
  in Mexico (1995), 105
  postwar, 48
  US (2000–2001), 5–6, 87, 112,
    138, 141, 144, 149, 155
religion, 2, 7, 11, 14–15, 17, 19,
  29–30, 43, 45, 50–1, 59,
  64, 106, 109–11, 126,
  127, 150
remembered utility, 26–8, 42
Ricardo, David, 99
risk aversion, 29, 163n8
Robeyns, I., 11–12, 15, 18, 56,
  59, 64–5
Robinson, 97–8, 125, 164n9
Rodgers, Y., 123

rule-utilitarianism, 22–3
Russia, 20, 51, 83
Russian-Polish Jews, 20, 51
Ruwanpura, K. N., 50

Sargent, J., 136, 143
Sawyer, Malcolm, 95, 99
Schmidt, K. M., 36–7, 162n2
Schönberg, U., 46
self-esteem, 42, 72
self-evaluation, 17, 19–20, 61,
    119–20
self-reported well-being, 24, 27–8,
    43, 61, 84
selves, 40–1
    See also multiple identities
Sen, Amartya K., 1–3, 9–13, 15–16,
    19–20, 31, 43–4, 53, 55,
    60–2, 67–9, 73, 80, 86, 119,
    161n1,2, 162n6,9,1,3
"set point of happiness," 2, 25–6,
    30, 69, 83–5, 89, 161n5
Shafir, E., 90
Shapiro, C., 94
"shirking" approach, 37, 94
Shoham, Y., 35
simultaneity, 96–9
Singer, P., 16, 22
slavery, 64
Smith, Adam, 10
Smith, R. S., 96
Smith, V. L., 13, 29, 40
"social capabilities," 31, 64–5, 79
social constructs, 3, 16, 31, 40, 81
social entitlements, 4, 7, 53–4,
    58–61, 65–6, 68, 71–2, 78,
    82, 88, 91, 98, 102–3,
    105–33, 159
    and cultural identity, 106–11
    and external shocks, 111–19
    and gender identity, 119–32
social identity theory, 42
social "law," 35
social norms, 1–2, 7, 9–10, 28,
    30–2, 33–5, 38–40, 42, 49,
    51–2, 55–6, 58–9, 66, 73–4,

78, 90, 94, 102, 105, 119,
    125–6, 131–2, 136, 138,
    141–3, 149–55, 158–9, 161n1,
    162n2,3
    and maquiladora productivity,
    151–5
Sonora, 112–15
Soskice, D., 94
South Africa, 19, 83
South Korea, 6
Southern-Italian Catholic, 51
Soviet Union, 83
Spain, 111
Spaniards, 107–9
status, 25, 39, 43, 60, 73, 84–6,
    100, 158, 162n7, 165n11
Steiner, H., 57–8
sterilization, 51
sticky wages, 34, 99–100
Stiglitz, J., 94
stratification economics, 39–40
Strober, M. H., 143–4
Stutzer, A., 89
subjective capabilities, 12–17
subjectivity in well-being, 2, 7,
    9–32, 43, 50–1, 62–3, 69, 75,
    82, 89, 99, 132, 159
    and capabilities, 12–21
    and capability
    approach, 10–12
    and happiness approach, 21–8
    and interdependence and
    adaptation, 28–32
Suh, E. M., 26
suicide, 18
Sunder, S., 40
supply and demand, 39,
    96–9, 143–4
survival, 13, 29

Taiwan, 43, 123
Tajfel, H., 42
Tamaulipas, 112–13, 115
Ten Commandments, 45
Tennenholtz, M., 35
Tenochtitlan, 107, 109

Texas, 5
time, 2, 4, 20–1, 24–7, 30, 31, 35–6, 39, 42–4, 49, 53–4, 59, 63–72, 77–85, 91–7, 126, 129, 150–8, 163n7
"treadmill effect," 28–9, 81
Turner, J., 42
Tversky, A., 31, 69, 161n5, 165n10

Ukraine, 83
underemployment, 122–4
UNDP, *See* United Nations Development Program
unemployment, 39, 46–7, 86, 91, 93–5, 99, 106, 120, 122–4, 164n14
UNICEF, 19
United Kingdom (UK), 46
United Nations (UN), 22
United Nations Development Program (UNDP), 12, 17, 125
Urry, H., 27
US Constitution, 14
US labor market, 46–9, 133
US recession (2000–2001), 5–6, 87, 112, 138, 141, 144, 149, 155
utilitarianism, 2, 10, 16, 21–3, 31, 58, 62
    act-utilitarianism, 22
    and happiness approach, 21–3
    hedonistic utilitarianism, 23
    rule-utilitarianism, 22
utility, 26–8, 42–3
    experienced utility, 26–8, 42–3
    maximization, *See* utility maximization
    remembered utility, 26–8, 42
utility maximization, 31, 34, 45, 69, 84, 157
Uyan-Semerci, P., 17, 31

Van Staveren, I., 16, 161n2
Veenhoven, R., 56
Vietnam, 88
Virgin of Guadalupe, 109–10
Vizard, P., 12, 16

wage-employment relationship, 92–102
    causality of, 95–6
    sign of, 92–4
wages, 34, 37–40, 45–6, 49, 59–60, 67–8, 75, 79, 84, 87–103, 120–2, 124–6, 136–44, 147–55, 158–9, 164n13,14, 165n6
    efficiency wages, 37
    and employment, *See* wage-employment relationship
    and fair wages, 38, 49, 94, 141–2, 149, 150, 155
    and free-rider behavior, 34
    and *maquiladora* industry, *See* wages and *maquiladora* industry
    and productivity, 151–5
    real wage, *See* real wage
    sticky wages, *See* sticky wages
    wage-employment relationship, 92–102
wages and *maquiladora* industry, 92–6, 139, 148–54
    hourly real-wages, 152
    log of the gender wage gap, 148
    log of productivity and real wages, 152
    nominal wages by gender, 139
    trends in the gender wage gap, 149–51
    variance decomposition of, 153
    wage-employment relationship, 92–6
Wang, Y., 46
Washington, D. C., 70–1
Weber, M., 14
well-being, 1–3, 6–7, 9–13, 17–31, 43, 50–1, 53, 55, 61–3, 65, 69, 75, 80, 82, 84, 86, 89, 105, 127, 129–32, 158–9
    self-reported, 24, 27–8, 43, 84
    subjective, *See* subjectivity in well-being
    research, 1–2

Wolf, E. R., 109–10
World Bank, 82–3
World Trade Organization (WTO),
    5, 87, 136, 141, 143–4,
    155, 164n3
World Value Survey, 82–3
World Wildlife Fund (WWF), 22
WTO, *See* World Trade
    Organization

Wvalle-Vazquez,
    Karina, 136
WWF, *See* World Wildlife
    Fund

Yellen, J. L., 94, 141, 165n6
Yucatan, 107

Zimbabwe, 20